COBIT® 2019

Implementing the NIST Cybersecurity Framework Using COBIT® 2019

ISACA®

About ISACA

Now in its 50th-anniversary year, ISACA® (*isaca.org*) is a global association helping individuals and enterprises achieve the positive potential of technology. Today's world is powered by information and technology, and ISACA equips professionals with the knowledge, credentials, education and community to advance their careers and transform their organizations. ISACA leverages the expertise of its 460,000 engaged professionals—including its 140,000 members—in information and cybersecurity, governance, assurance, risk and innovation, as well as its enterprise performance subsidiary, *CMMI® Institute*, to help advance innovation through technology. ISACA has a presence in more than 188 countries, including more than 220 chapters worldwide and offices in both the United States and China.

Disclaimer

Reservation of Rights

ISACA
1700 E. Golf Road, Suite 400
Schaumburg, IL 60173, USA
Phone: +1.847.660.5505
Fax: +1.847.253.1755
Contact us: https://support.isaca.org
Website: www.isaca.org

Participate in the ISACA Online Forums: https://engage.isaca.org/onlineforums

Twitter: http://twitter.com/ISACANews
LinkedIn: www.linkedin.com/company/isaca
Facebook: www.facebook.com/ISACAGlobal
Instagram: www.instagram.com/isacanews/

Implementing the NIST Cybersecurity Framework Using COBIT® 2019
ISBN 978-1-60420-815-3

Acknowledgments

Expert Reviewers
Tom Conkle, CISSP, Optic Cyber Solutions, USA
Kelly Hood, CISSP, Optic Cyber Solutions, USA
Mark Thomas, CRISC, CGEIT, Escoute, USA
Greg Witte, CISM, USA

Page intentionally left blank

TABLE OF CONTENTS

LIST OF FIGURES

Chapter 1. Introduction

Chapter 2. Introduction to NIST Cybersecurity Framework V1.1

Chapter 3. Framework Implementation

Appendix B. Detailed Template for Framework Profile

Appendix C. Action Planning

Page intentionally left blank

Executive Summary

Information technology (IT) and operational technology (OT) do not merely support key information resources across a wide variety of enterprises today. On the contrary, IT and OT—along with the information assets they curate—*increasingly constitute the core of the business itself*, create new value and represent the most attractive areas of opportunity for the foreseeable future.

Information and technology (I&T), their associated disciplines, and I&T-related infrastructure and processes have become increasingly advanced, pervasive and interconnected, which expands the attack surface. Accordingly, devastating assaults on financial, retail and energy providers (to name just a few examples) required enterprises across the spectrum to rethink their management of I&T-related risk to maintain acceptable levels.

Many enterprises recognize the challenge but lack an approach that integrates cybersecurity standards and enterprise governance of I&T (EGIT) to establish systematic—yet flexible and achievable—governance and management objectives, processes and capability levels to make measured improvements toward cybersecurity goals. This publication describes proven practices to anticipate, understand and optimize I&T risk by implementing the US National Institute of Standards and Technology (NIST) *Framework for Improving Critical Infrastructure Cybersecurity*[1] using COBIT® 2019. Applying the NIST Cybersecurity Framework (CSF) in the context of COBIT 2019 enhances communication about priorities for I&T, helps to maximize I&T for competitive advantage and brings transparency to the definition and management of I&T risk, all in terms of the business and its key goals.

The Cybersecurity Enhancement Act of 2014[2] (CEA) updated the role of NIST to include ongoing development and support of a voluntary, consensus-based, industry-led and cost-effective set of standards, guidelines, best practices, methodologies, procedures and processes to reduce cyberrisk to critical infrastructure. Under CEA, NIST extended previous work on Cybersecurity Framework Version 1.0, initiated in February 2013 by Executive Order (EO) 13636, "Improving Critical Infrastructure Cybersecurity." NIST received comments from US and international organizations—including ISACA®—and released CSF Version 1.1 in April 2018.

The CSF continues to provide a risk-based approach to cybersecurity maturity that emphasizes business drivers and supports flexibility, customization and broad applicability across public and private sectors alike. The CSF closely mirrors enterprise I&T governance and management principles that ISACA has fostered for many years—the very concepts that formed the ISACA contribution to CSF Version 1.1. Thus, ISACA COBIT 2019 natively supports implementation of the CSF while leveraging the lessons learned in over 50 years of ISACA success.

This publication maps CSF steps and activities to COBIT 2019, correlating CSF guidance with measurable governance and management practices.

COBIT 2019 helps enterprises to achieve CSF objectives by:

- Identifying stakeholder needs and high-level enterprise goals
- Translating enterprise goals into specific governance and management objectives, and ultimately into prioritized alignment goals for I&T and associated processes, practices and activities
- Developing a tailored, nuanced understanding of I&T risk within specific business contexts
- Applying customizable design factors that help enterprises to identify, balance and integrate critical I&T initiatives for maximum business impact—while avoiding a disconnected checklist model of IT transformation that can lead to diffuse and/or piecemeal implementation
- Discerning target capability levels for cybersecurity processes and assessing their achievement

[1] NIST, *Framework for Improving Critical Infrastructure Cybersecurity*, Version 1.1, USA, 16 April 2018, *https://nvlpubs.nist.gov/nistpubs/CSWP/NIST.CSWP.04162018.pdf*

[2] Library of Congress, "S.1353 - Cybersecurity Enhancement Act of 2014," Public Law No. 113-274, *https://www.congress.gov/bill/113th-congress/senate-bill/1353/text*

- Informing selection, priority and implementation of key security controls referenced by CSF, using concepts such as the goals cascade and COBIT® implementation roadmap

The CSF was originally intended to support critical infrastructure providers; however, any enterprise striving to manage and reduce cybersecurity risk can benefit from it. Nearly all enterprises, in some way, are part of critical infrastructure. Each enterprise is connected to critical functions as a consumer through the global economy, telecommunication services and many other ways. Improvements in risk management by each member of this ecosystem cumulatively reinforces the efforts of others, thereby reducing cybersecurity risk globally.

A key contributor to CSF development and an active participant in national cybersecurity workshops, ISACA brings a unique and valuable perspective to implementation of the Cybersecurity Framework. The native expertise of ISACA inspires all guidance and templates in this publication that guides enterprises to attain CSF outcomes in a systematic and measurable, yet adaptable, way; enumerate risk (and potential impact); anticipate unforeseen circumstances; minimize losses and gain business advantage.

Chapter 1
Introduction

1.1 Background

Threats to information security systems are not new. ISACA was established 50 years ago to address the need for a centralized source of information and guidance for securing computer systems. Today's cybersecurity attacks have evolved well beyond defacement of webpages and viruses. In more recent years, mass data breaches of personally identifiable information (PII) and pervasive theft of intellectual property (IP) span continents, and cybersecurity incidents now routinely entail denial, degradation and/or disruption of critical infrastructure and other governmental functions and information. Attackers are more organized, well-supported and leverage sophisticated methods that dwarf the achievements of bad actors of just 20 years ago. At the same time, society is growing ever more dependent on technology, and connectivity and information sharing have become increasingly vital for citizens, educators, business professionals, corporations, institutions and countries—in short, for everyone. As digital transformation continues to proliferate and high velocity I&T continue to evolve, the need to protect against cybersecurity attacks is more critical than ever.

Today's cybersecurity attacks have evolved well beyond defacement of webpages and viruses. In more recent years, mass data breaches of personally identifiable information (PII) and pervasive theft of intellectual property (IP) span continents, and cybersecurity incidents now routinely entail denial, degradation and/or disruption of critical infrastructure and other governmental functions and information.

To help address these needs, ISACA launched a security knowledge platform and cybersecurity professional program in 2014. Developed in collaboration with cybersecurity experts from leading enterprises around the world, Cybersecurity Nexus™ (CSX) supplies cutting-edge thought leadership, training and credentialing for professionals at all career stages who are leading cybersecurity into the future. Adding to its growing information security portfolio in 2014, ISACA published its original *Implementing the NIST Cybersecurity Framework*, a resource companion for the US National Institute of Standards and Technology (NIST) Cybersecurity Framework (CSF). *Implementing the NIST Cybersecurity Framework Using COBIT® 2019* updates the original ISACA publication with COBIT 2019, the latest edition of the ISACA classic, comprehensive framework for enterprise governance of information and technology (EGIT).

Although NIST originally created the CSF to support critical infrastructure providers, it can be applied to any enterprise that incurs cybersecurity risk, i.e., all businesses, organizations and entities that participate in or support critical infrastructure. All enterprises are connected to—and depend on—critical functions and services as consumers, through the global economy, telecommunication services and many other ways. Improved risk management by each implementer will ultimately lower risk levels for all—with a cumulative effect of reducing global cybersecurity risk.

This publication addresses business and technical requirements to apply the CSF, leveraging COBIT 2019. The intended audience ranges from board members and executive management to operations and maintenance professionals. **Figure 1.1** identifies several key roles/functions and the benefit that each receives from the CSF. Although these roles are generally applicable across all sectors and industry, specific tailoring may be required to correlate, align and/or adapt the roles within given organizational contexts.

This publication addresses business and technical requirements to apply the CSF, leveraging COBIT 2019.

Figure 1.1—CSF Implementation: Target Audience and Benefits

Framework Role	Audience Role/Function	Benefit/Reason for Applying the Framework
Executive	Board and executive management	• Understanding scope of responsibility and roles with respect to cybersecurity enterprisewide • Understanding current cybersecurity posture • Understanding cybersecurity risk to the organization • Understanding cybersecurity target state • Understanding actions required to close gap between current cybersecurity posture and target state
Business/Process	IT management	• Awareness of business impact • Understanding relationship of business systems to risk appetite
Business/Process	IT process management	• Understanding business requirements, mission objectives and priorities
Business/Process	Risk management	• Enhanced view of operational environment to discern likelihood of cybersecurity events
Business/Process	Legal experts	• Understanding cyberthreats to business units and their mission objectives • Understanding all compliance requirements for each business unit
Implementation/ Operations	Implementation team	• Understanding security controls and their importance in managing operational security risk • Detailed understanding of actions necessary to close gaps in cybersecurity requirements
Implementation/ Operations	Employees	• Understanding cybersecurity requirements for relevant business systems

1.2 Governance and Management of Enterprise Information and Technology (I&T)

ISACA is dedicated to enhancing the knowledge and skills that practitioners need to set and achieve strategic goals and realize business benefits through effective and innovative use of technology. This publication uses the following terms to describe key contexts, plans, processes and activities:

- **Enterprise**—A group of individuals working together for a common purpose, typically within the context of an organizational form, such as a corporation, partnership, limited liability company, public agency, charity or trust
- **Organization**—The structure or arrangement of related or connected components of an enterprise, defined by a particular scope
- **Governance**—The framework and system ensuring that:
  - Stakeholder needs, conditions and options are evaluated to determine balanced, agreed-on enterprise objectives
  - Strategic direction is set, and goals are prioritized and supported through appropriate, timely decision making
  - Performance and compliance are monitored against agreed-on enterprise strategy and objectives
- **Management**—Planning, building, operating and monitoring activities, in alignment with the direction set by the governance body, to achieve enterprise objectives

COBIT 2019 regularly references information and technology (I&T). In the context of this guide, I&T refer to technical processes and solutions (hardware and software), data, and information across the enterprise that enable business functions to achieve enterprise objectives. It is important to note that, in this context, technology includes not only operational technology (OT) (e.g., industrial control systems and building management systems), but also traditional information technology (IT) (e.g., payroll and email systems). I&T include all the information and technology that the enterprise generates, processes and uses to achieve its goals, regardless of where that information and technology reside in the enterprise (logistically or organizationally). Enterprise I&T are not limited to the IT or OT department of an enterprise, but include the technologies, data and processes traditionally associated with those disciplines and/or organizational departments.

Technical systems continue to converge, and the systems that enable enterprise value are increasingly interconnected. The programmable logic controllers that support a manufacturing process, for example, use computing devices similar to those that support office printing needs—and both require effective cybersecurity governance and management.

Throughout this guide, the reader is encouraged to adopt a comprehensive view of technology. The planning and management processes described here may help enterprises to evaluate and support convergence of OT and IT. For example, determining critical organizational resources uses COBIT 2019 practice APO07.02 *Identify key IT personnel*. This practice often includes consideration of an important database administrator or data center operator, but should be expanded to consider personnel who maintain important industrial control system (ICS) components, physical security systems (e.g., access control systems and cameras) and/or facility operations (phone, HVAC [heating, ventilation and air conditioning], electrical and emergency systems). A broad view of enterprise technology helps support effective cybersecurity management in all planning, building, operating and monitoring activities.

1.3 Background of the *Framework for Improving Critical Infrastructure Cybersecurity*

Executive Order (EO) 13636, "Improving Critical Infrastructure Cybersecurity," was issued to protect the United States from cybersecurity attacks that could disrupt power, water, communication and other critical systems.[3] The EO directs the executive branch of the US government to collaborate with industrial partners around the world to:[4]

- Promote and incentivize adoption of cybersecurity practices
- Increase volume, timeliness and quality of cyberthreat information sharing
- Incorporate strong privacy and civil-liberties protections into every initiative to secure critical infrastructure
- Explore the use of existing regulation to promote cybersecurity

Additionally, Presidential Policy Directive (PPD) 21, *Critical Infrastructure Security and Resilience,* directs the executive branch of the US government to:

- Develop a situational-awareness capability that comprehends physical and cyber dimensions of critical infrastructure and its functioning in near real time
- Understand the cascading consequences of infrastructure failures
- Evaluate and mature public-private partnerships
- Update the National Infrastructure Protection Plan
- Develop a comprehensive research and development plan

Figure 1.2 identifies critical sectors that are addressed in PPD-21 and indicates sector-specific agencies (SSAs) with designated responsibility.

[3] Executive Order 13636 is available from the US Government Printing Office at *www.gpo.gov/fdsys/pkg/FR-2013-02-19/pdf/2013-03915.pdf.*

[4] Information related to EO 13636 is drawn from the US Department of Homeland Security's fact sheet for EO 13636 and PPD-21, available at *https://www.dhs.gov/sites/default/files/publications/eo-13636-ppd-21-fact-sheet-508.pdf*; for full text of Presidential Policy Directive (PPD) 21, see The White House: President Barack Obama, Statements & Releases, "Presidential Policy Directive — Critical Infrastructure Security and Resilience," *https://obamawhitehouse.archives.gov/the-press-office/2013/02/12/presidential-policy-directive-critical-infrastructure-security-and-resil.*

Figure 1.2—Sector-specific Agencies Designated in PPD-21

Sector	Sector-Specific Agency or Agencies
Chemical	Department of Homeland Security
Commercial Facilities	Department of Homeland Security
Communications	Department of Homeland Security
Critical Manufacturing	Department of Homeland Security
Dams	Department of Homeland Security
Defense Industrial Base	Department of Defense
Emergency Services	Department of Homeland Security
Energy	Department of Energy
Financial Services	Department of the Treasury
Food and Agriculture	Departments of Agriculture and Health and Human Services
Government Facilities	Departments of Homeland Security and General Services Administration
Healthcare and Public Health	Department of Health and Human Services
Information Technology	Department of Homeland Security
Nuclear Reactors, Materials and Waste	Department of Homeland Security
Transportation Systems	Departments of Homeland Security and Transportation
Water and Wastewater Systems	Environmental Protection Agency

Section 7 of EO 13636 directed the Secretary of Commerce to task NIST with leading the development of a framework (the Cybersecurity Framework) to reduce cyberrisk to critical infrastructure. The CSF should include a set of standards, methodologies, procedures and processes that align policy, business and technological approaches to address cyberrisk. The EO further directed NIST to incorporate voluntary consensus standards and industry best practices, and to be consistent with voluntary international standards when such international standards will advance the objectives of the EO.

Success criteria for the CSF reside in section 7 of EO 13636, and require that the CSF:

- Provide a prioritized, flexible, repeatable, performance-based and cost-effective approach, including information security measures and controls, to help owners and operators of critical infrastructure identify, assess and manage cyberrisk
- Focus on identifying cross-sector security standards and guidelines applicable to critical infrastructure
- Identify areas for improvement that should be addressed through future collaboration with particular sectors and standards-developing organizations
- Provide technology-neutral guidance that enables critical infrastructure sectors to benefit from a competitive market for products and services meeting the standards, methodologies, procedures and processes developed to address cyberrisk.
- Include guidance for measuring the performance of an entity in implementing the Cybersecurity Framework

In response to this directive, NIST issued a request for information (RFI) in February 2013, asking a broad array of questions to gather relevant input from industry, academia and other stakeholders. NIST solicited information regarding:

- How enterprises assess risk
- How cybersecurity factors into risk assessment
- To what extent enterprises adopt existing cybersecurity frameworks, standards, guidelines and other management practices related to cybersecurity

In addition, NIST asked about legal and regulatory aspects of particular frameworks, standards, guidelines and best practices, and the challenges that enterprises perceive in meeting such requirements.

NIST subsequently conducted five workshops throughout the United States to refine industry feedback, which included significant contributions from ISACA and its membership. Based on responses to the RFI and results from the workshops, NIST provided a Cybersecurity Framework that identifies existing practices to inform risk management decisions related to the prevention and detection of, response to, and recovery from cybersecurity issues.

The Cybersecurity Enhancement Act of 2014[5] (CEA) updated the role of NIST to include ongoing development and support of a voluntary, consensus-based, industry-led and cost-effective set of standards, guidelines, best practices, methodologies, procedures and processes to reduce cyberrisk to critical infrastructure. CEA provided guidance that enabled NIST to revise CSF Version 1.0, released 12 February 2014. CSF Version 1.1 was released 16 April 2018 incorporating feedback from US and international sectors (including ISACA). Version 1.1 continues to include the three primary CSF components: Framework Core, Framework Implementation Tiers and Framework Profiles—all discussed in further detail in chapter 2 in this publication.

1.4 Introduction to COBIT 2019

COBIT 2019 provides a comprehensive framework that assists enterprises in achieving their objectives for enterprise governance of I&T (EGIT). COBIT 2019 can be implemented incrementally—starting small and building on initial success—or holistically for the entire enterprise, encompassing the full, end-to-end business and I&T functional areas of responsibility. In either approach, COBIT helps enterprises create optimal value from I&T by balancing business benefit with risk and optimizing resource use. COBIT 2019 incorporates an expanded definition of governance and has been streamlined for flexible adoption—whether for discrete projects or entire enterprises.

The COBIT 2019 product family is open-ended and designed for customization; forthcoming titles will supplement and elaborate the core product. Current COBIT 2019 publications and resources are listed at *http://www.isaca.org/COBIT/Pages/COBIT-2019-Publications-Resources.aspx*.

COBIT 2019 reflects two sets of foundational concepts:

- Principles for a governance system that describe the core requirements of a governance system for enterprise information and technology (I&T)
- Principles for a governance framework that can be used to build a governance system for the enterprise

Principles for a governance system include (**figure 1.3**):

1. Providing stakeholder value
2. Approaching governance holistically
3. Viewing governance as a dynamic system
4. Distinguishing governance from management
5. Tailoring governance to enterprise needs
6. Designing for the end-to-end governance system

[5] *Op cit* Library of Congress

Figure 1.3—Governance System Principles

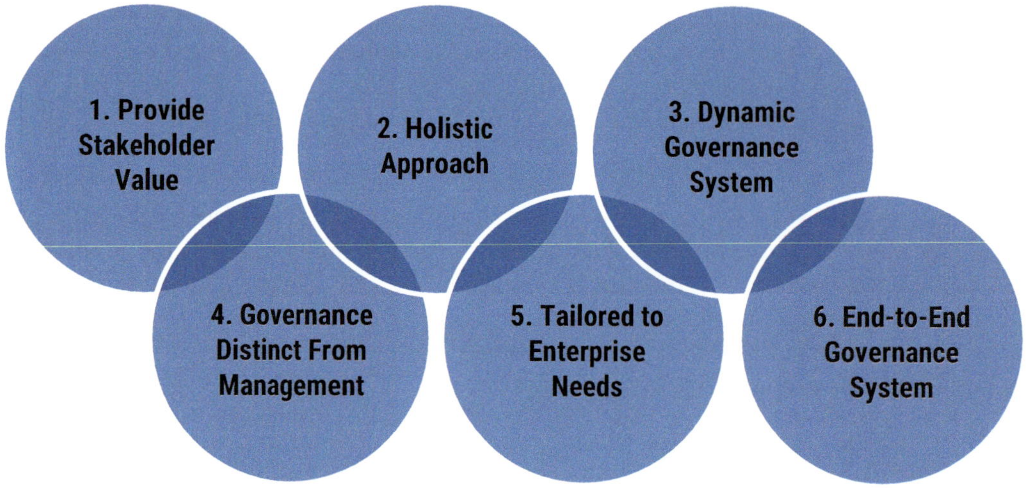

Source: ISACA, *COBIT® 2019 Framework: Introduction and Methodology*, USA, 2018, *https://www.isaca.org/COBIT/Pages/COBIT-2019-Framework-Introduction-and-Methodology.aspx*

Principles for a governance framework require that it be (**figure 1.4**):

1. Based on a conceptual model
2. Open and flexible
3. Aligned to major standards

Figure 1.4—Governance Framework Principles

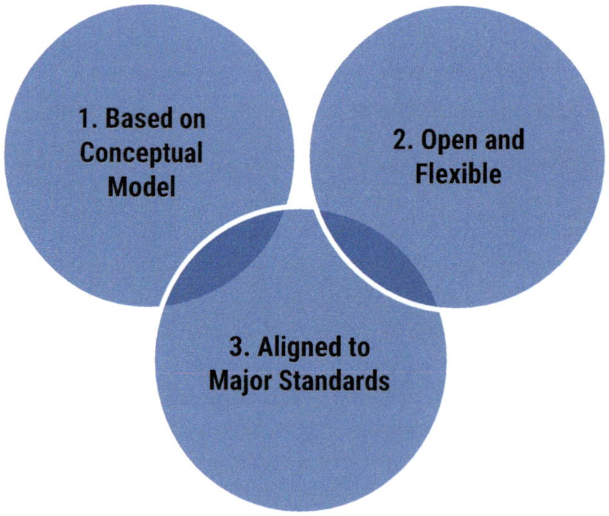

Source: ISACA, *COBIT® 2019 Framework: Introduction and Methodology*, USA, 2018

Together, these principles enable an enterprise to build an effective governance and management framework that optimizes information and technology investment for the benefit of stakeholders.

Enterprises exist to create value for stakeholders. Consequently, any enterprise will adopt value creation as a prime governance objective (**figure 1.5**). Value creation means delivering value from digital transformation and mitigating any related business risk. Creating value through proper governance means that the enterprise realizes benefits at an optimal resource cost while optimizing risk and aligning I&T with enterprise goals. Benefits can take many forms, e.g., financial profit for private, commercial enterprises, and taxpayer benefits or public services for government entities.

Figure 1.5—The Context of Enterprise Governance of Information and Technology

Source: De Haes, Steven; W. Van Grembergen: *Enterprise Governance of Information Technology: Achieving Alignment and Value, Featuring COBIT 5, 2nd ed.*, Springer International Publishing, Switzerland, 2015, *https://www.springer.com/us/book/9783319145464*

COBIT 2019 Distinguishes Governance From Management

The COBIT 2019 framework makes a clear distinction between governance and management. These two disciplines encompass different types of activities, require different organizational structures and serve different purposes:

- **Governance**—Governance ensures that:
  - Performance and compliance are monitored against agreed-on enterprise strategy and objectives
  - Strategic direction is set, and goals are prioritized and supported through appropriate, timely decision making
  - Performance and compliance are monitored against agreed-on enterprise strategy and objectives
- **Management**—Management plans, builds, runs and monitors activities in alignment with the direction set by the governance body to achieve enterprise objectives.

COBIT 2019 Goals Cascade

Enterprise stakeholder needs must be transformed into an actionable strategy. The COBIT 2019 goals cascade (**figure 1.6**) allows practitioners to translate stakeholder needs into specific, actionable and customized enterprise goals, alignment goals and governance and management objectives. The goals cascade yields specific goals at every organizational level, in each area of the enterprise, localizing and disseminating the overall, enterprisewide goals and stakeholder requirements, and thus effectively supports alignment between enterprise needs and IT solutions and services.

Figure 1.6—COBIT Goals Cascade

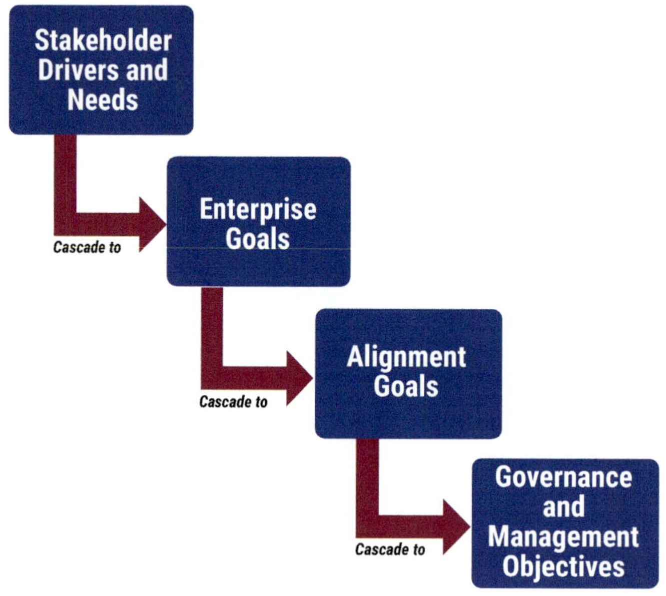

Source: ISACA, *COBIT® 2019 Framework: Introduction and Methodology*, USA, 2018

COBIT 2019 Components of the Governance System

COBIT 2019 provides a holistic and systemic view of EGIT, based on several components. Components reflect factors that contribute individually and collectively to the good operations of the enterprise governance system over I&T. Components are driven by the goals cascade, which ensures that the enterprise higher-level goals define what the different components of the system should achieve.

The COBIT 2019 framework describes seven governance components (**figure 1.7**):

1. Processes
2. Organizational structures
3. Principles, policies and procedures
4. Information
5. Culture, ethics and behavior
6. People, skills and competencies
7. Services, infrastructure and applications

Figure 1.7—COBIT Components of a Governance System

Source: ISACA, *COBIT® 2019 Framework: Introduction and Methodology*, USA, 2018

Components interact with each other, resulting in a holistic governance system for I&T. Components of all types can be generic or can be variants of generic components:

- **Generic** components are described in the COBIT core model (**figure 1.8**) and apply in principle to any situation. However, they are generic in nature and generally need customization before being practically implemented.
- **Variants** are based on generic components but are tailored for a specific purpose or context within a focus area (e.g., for information security, DevOps or a particular regulation).

COBIT 2019 Core Model

Each governance and management objective relates to one process (with an identical or similar name). A process is one of seven components that work together to achieve governance and management objectives. Each process includes an organized set of practices and activities to achieve certain objectives and produce a set of outputs that support achievement of overall IT-related goals.

The COBIT core model (**figure 1.8**) captures the 40 governance and management processes defined in detail in COBIT 2019. The core model represents all the processes relating to I&T activities normally found in an enterprise, offering a common reference model that is accessible to both operational IT and business managers. The core model is complete and comprehensive—however, it is not the only possible process model. Each enterprise must define its own process set, considering its own specific situation.

Figure 1.8—COBIT Core Model

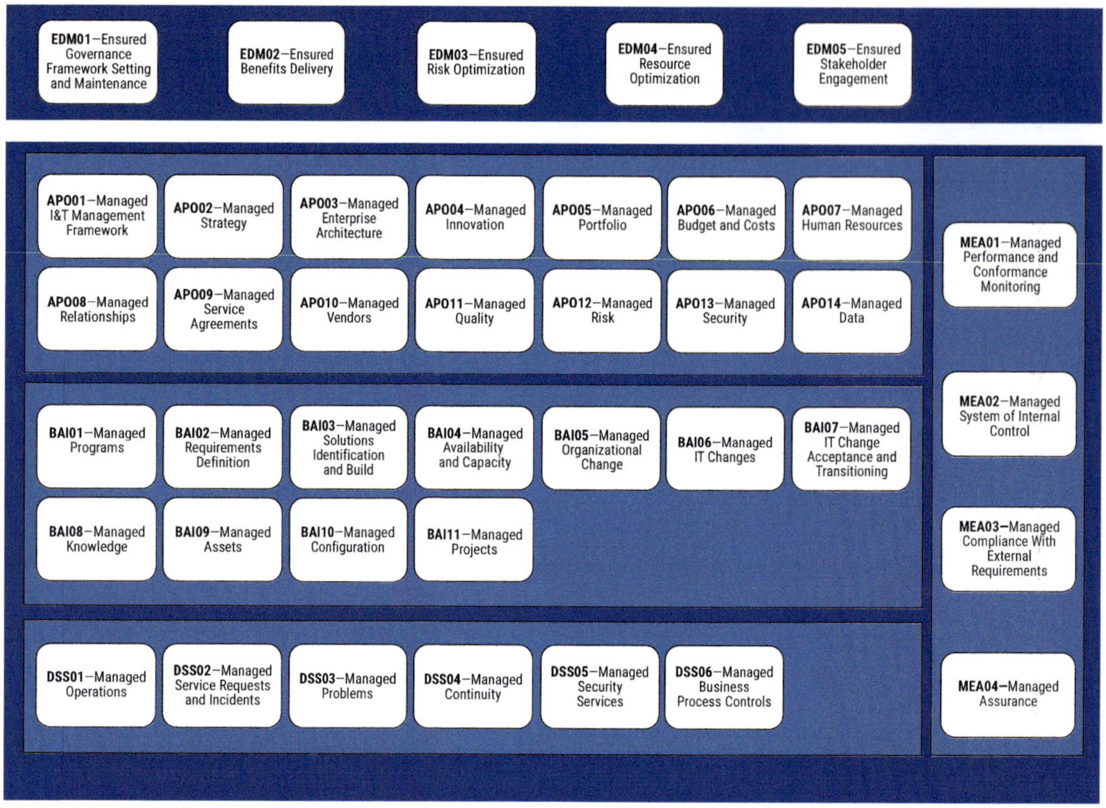

Source: ISACA, *COBIT® 2019 Framework: Introduction and Methodology*, USA, 2018

Adopting an operational model and a common language for all parts of the enterprise involved in I&T activities is one of the most important and critical steps toward good governance. This step also provides a framework for measuring and monitoring I&T performance, communicating with service providers and integrating best management practices.

Performance Management in COBIT

Developed on the basis of concepts and methodology that distinguish several tiers of capability and maturity levels, COBIT 2019 performance management (CPM) is designed to evaluate how well the governance and management system and all the components of an enterprise work, and how they can be enhanced to the required level. CPM is intended to be:

1. Simple to understand and use
2. Consistent with and supportive of the COBIT conceptual model
3. Reliable, repeatable and relevant
4. Flexible
5. Suited to multiple types of assessments

The CPM model largely aligns with and extends CMMI© V2.0 concepts:

- Process activities are associated with target capability levels, as specified in the detailed guidance of chapter 4, *COBIT® 2019 Framework: Governance and Management Objectives*.
- Other components (e.g., organizational structures and information) may also have capability levels defined in future guidance that ISACA may release.
- Maturity levels are associated with focus areas (i.e., a collection of governance and management objectives and underlying components) and are achieved if all required capability levels are achieved.

COBIT 2019 Design Guide

To help practitioners determine where to start and realize the greatest return on investment of time and resources, ISACA created the *COBIT® 2019 Design Guide: Designing an Information and Technology Governance Solution*. This guide explores design factors that can influence governance and includes a workflow for tailoring a governance system that positions any enterprise for success in the use of I&T.

COBIT 2019 Design Factors

Enterprises differ in many respects, for example, size, sector, regulatory landscape, threat landscape, role of IT for the enterprise and tactical technology-related choices (**figure 1.9**). COBIT 2019 collectively refers to these aspects as design factors. Thoughtful application of these parameters enables any enterprise to tailor its governance system and derive the most value out of I&T.

Tailoring means that an enterprise starts from the COBIT core model and applies changes to the generic framework based on the relevance and importance of a series of design factors. This process is called "designing the governance system for enterprise information and technology."

Figure 1.9—COBIT Design Factors

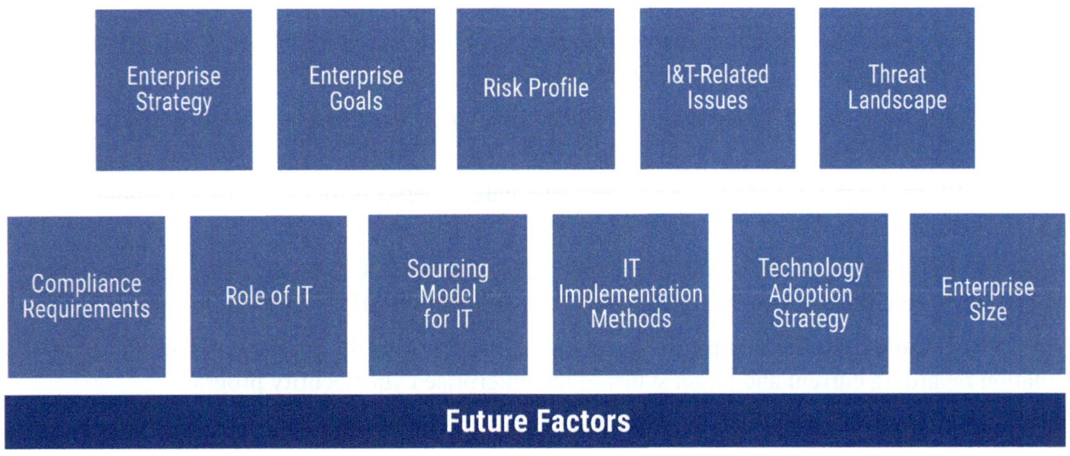

Source: ISACA, *COBIT® 2019 Framework: Introduction and Methodology*, USA, 2018

COBIT 2019 Implementation Guidance

Practitioners realize optimal value from COBIT when they adapt the framework effectively to suit the unique environment of a given enterprise. Each implementation should address specific challenges, including managing changes to culture and behavior.

ISACA provides extensive, practical implementation guidance in its publication, *COBIT® 2019 Implementation Guide: Implementing and Optimizing an Information and Technology Governance Solution,* which is based on a continual improvement life cycle. This guide is not intended to be a prescriptive approach nor a complete solution, but rather a guide to avoid commonly encountered pitfalls, leverage good practices and assist in the creation of successful outcomes.

The following are important topics covered in the *COBIT® 2019 Implementation Guide*:

- Positioning enterprise governance of I&T within the enterprise internal and external environment(s)
- Recognizing typical pain points and trigger events
- Creating the appropriate environment for implementation
- Leveraging COBIT to identify gaps and guide the development of components, such as policies, processes, principles, organizational structures, and roles and responsibilities

Where appropriate, the *COBIT® 2019 Design Guide*[6] may be beneficial because it elaborates on certain tasks in the *COBIT® 2019 Implementation Guide.*

1.5 Scope and Approach

The guidance in this publication is intended to assist enterprises with CSF implementation using ISACA methodologies, especially those developed in COBIT® 2019. The guide includes processes, example templates, and strategic and tactical direction designed to maximize benefit from the CSF and to help practitioners identify and achieve enterprise objectives for the governance and management of I&T.

The information is organized as follows:

- **Chapter 2**—Provides a detailed introduction into the NIST Cybersecurity Framework V1.1 and its three components: Framework Core, Implementation Tiers and Profiles
- **Chapter 3**—Describes an optimal approach—along with supporting templates—for implementing the CSF to improve EGIT holistically
- **Chapter 4**—Illustrates the use of the CSF to communicate cybersecurity requirements among internal and external stakeholders
- **Appendix A: Framework Core**—Provides a copy of the Framework Core for quick reference that is updated throughout with new Informative References that are mapped to COBIT® 2019 elements
- **Appendix B: Detailed Profile Template**—Provides an overview of the profile template that is used to collect information regarding current and target states of the enterprise cybersecurity program
- **Appendix C: Action Planning**—Provides considerations for developing an action plan to track gap-closing measures
- **Appendix D: Considerations for Critical Infrastructure Sectors**—Provides considerations and priorities for tailoring CSF implementation

[6] ISACA, *COBIT® 2019 Design Guide: Designing an Information and Technology Governance Solution*, USA, 2018, *https://www.isaca.org/COBIT/Pages/COBIT-2019-Design-Guide.aspx*

Chapter 2
Introduction to NIST Cybersecurity Framework V1.1

2.1 Framework Background

The CSF was developed in response to US Presidential Executive Order 13636,[7] which states:

> *Repeated cyber intrusions into critical infrastructure demonstrate the need for improved cybersecurity. The cyber threat to critical infrastructure continues to grow and represents one of the most serious national security challenges we must confront.*

Although the executive order (EO) established the CSF, the Cybersecurity Enhancement Act of 2014 (CEA) codified the CSF into law. CEA is "an Act to provide for an ongoing, voluntary public-private partnership to improve cybersecurity, and to strengthen cybersecurity research and development, workforce development and education, and public awareness and preparedness, and for other purposes."[8] CEA defines the scope, authority and responsibilities regarding cybersecurity for a range of governmental entities that are identified in the Act.

The goals of the EO align well with the COBIT framework, which recognizes that "information and technology (I&T) have become crucial in the support, sustainability and growth of enterprises."[9] COBIT can optimize value from enterprise I&T by helping practitioners balance benefits with risk and resource consumption through holistic understanding, governance and management of all enterprise I&T processing and assets.

The CSF incorporates significant contributions from small- and medium-sized enterprises (SMEs), domestically in the United States and throughout the international business community. Their diverse input expanded coverage of the challenges and root causes underlying modern cybersecurity risk. Support from SMEs in particular contributed breadth and flexibility to the framework. NIST identifies recurring themes and considerations (**figure 2.1**) across critical infrastructure sectors and other industries:

- **Framework Principles**—Guiding concepts recommended by CSF contributors
- **Common Points**—Practices with wide utility and high rates of adoption among contributors
- **Initial Gaps**—Areas where responses from contributors were not sufficient to satisfy intention of executive order

Figure 2.1—NIST Initial Framework Considerations

Category	Framework Principles	Common Points	Initial Gaps
Themes	• Flexibility • Impact on global operations • Risk management approaches • Leverage existing approaches, standards and best practices	• Senior management engagement • Understanding threat environment • Business risk/risk assessment • Separation of business and operational systems • Models/levels of maturity • Incident response • Cybersecurity workforce	• Metrics • Privacy/civil liberties • Tools • Dependencies • Industry best practices • Resiliency • Critical infrastructure cybersecurity nomenclature

Source: NIST, *Initial Analysis of Cybersecurity Framework RFI Responses*, USA, 2013, figure 1, *https://www.nist.gov/file/368111*

[7] Executive Order (EO) 13636 is available from the US Government Printing Office at *www.gpo.gov/fdsys/pkg/FR-2013-02-19/pdf/2013-03915.pdf.*
[8] *Op cit* Library of Congress
[9] ISACA, *COBIT® 2019 Framework: Introduction and Methodology*, USA, 2018, *https://www.isaca.org/COBIT/Pages/COBIT-2019-Framework-Introduction-and-Methodology.aspx*

The CSF develops a risk-based approach to cybersecurity across three major components:

- Framework Core
- Framework Implementation Tiers
- Framework Profiles

Each CSF component reinforces the connection between business drivers and cybersecurity activities.

The **Framework Core** includes a set of cybersecurity activities, desired outcomes and Informative References that are common across critical infrastructure sectors.

Framework Implementation Tiers provide context on how an enterprise views cybersecurity risk and the processes in place to manage that risk. Tiers describe the degree to which cybersecurity risk management practices of an enterprise exhibit the characteristics defined in the Framework (e.g., risk- and threat-awareness, consistency and expertise, predictive capacity and continuous improvement). The Tiers characterize enterprise practices over a range, from Partial (Tier 1) to Adaptive (Tier 4).

A **Framework Profile** represents the outcomes based on business needs that an enterprise has selected from the Framework Core categories and subcategories. The Profile can be characterized as the alignment of standards, guidelines and practices to the Framework Core in the context of a particular implementation. Profiles also identify opportunities for improving cybersecurity posture by comparing a Current Profile (the as-is state) with a Target Profile (the future or aspired state).

In addition to providing a cybersecurity framework, the CSF also provides basic implementation guidance through a seven-step process. These CSF steps have been edited slightly to aid a global audience.

- **Step 1: Prioritize and Scope**—The organization identifies its business/mission objectives and high-level organizational priorities. With this information, the organization makes strategic decisions regarding cybersecurity implementations and determines the scope of systems and assets that support the selected business line or process. The Framework can be adapted to support the different business lines or processes within an organization, which may have different business needs and associated risk tolerance. Risk tolerances may be reflected in a target Implementation Tier.
- **Step 2: Orient**—Once the scope of the cybersecurity program has been determined for the business line or process, the organization identifies related systems and assets, regulatory requirements, and overall risk approach. The organization then consults sources to identify threats and vulnerabilities applicable to those systems and assets.
- **Step 3: Create a Current Profile**—The organization develops a Current Profile by indicating which Category and Subcategory outcomes from the Framework Core are currently being achieved. If an outcome is partially achieved, noting this fact will help support subsequent steps by providing baseline information.
- **Step 4: Conduct a Risk Assessment**—This assessment could be guided by the organization's overall risk management process or previous risk assessment activities. The organization analyzes the operational environment in order to discern the likelihood of a cybersecurity event and the impact that the event could have on the organization. It is important that organizations identify emerging risks and use cyber threat information from internal and external sources to gain a better understanding of the likelihood and impact of cybersecurity events.
- **Step 5: Create a Target Profile**—The organization creates a Target Profile that focuses on the assessment of the Framework Categories and Subcategories describing the organization's desired cybersecurity outcomes. Organizations also may develop their own additional Categories and Subcategories to account for unique organizational risks. The organization may also consider influences and requirements of external stakeholders such as sector entities, customers, and business partners when creating a Target Profile. The Target Profile should appropriately reflect criteria within the target Implementation Tier.

- **Step 6: Determine, Analyze, and Prioritize Gaps**—The organization compares the Current Profile and the Target Profile to determine gaps. Next, it creates a prioritized action plan to address gaps—reflecting mission drivers, costs and benefits, and risks—to achieve the outcomes in the Target Profile. The organization then determines resources, including funding and workforce, necessary to address the gaps. Using Profiles in this manner encourages the organization to make informed decisions about cybersecurity activities, supports risk management, and enables the organization to perform cost-effective, targeted improvements.
- **Step 7: Implement Action Plan**—The organization determines which actions to take to address the gaps, if any, identified in the previous step and then adjusts its current cybersecurity practices in order to achieve the Target Profile. For further guidance, the Framework identifies example Informative References regarding the Categories and Subcategories, but organizations should determine which standards, guidelines, and practices, including those that are sector specific, work best for their needs.

Hundreds of enterprises provided input into the design of the Cybersecurity Framework. ISACA was deeply engaged in the CSF development at each stage. As a result, many ISACA principles inform CSF implementation steps. **Figure 2.2** illustrates some parallels between the CSF and COBIT 2019 governance system principles.

Figure 2.2—Application of CSF to COBIT 2019 Governance System Principles

COBIT 2019 Governance System Principles	CSF Application
1. Each enterprise needs a governance system to satisfy stakeholder needs and to generate value from the use of I&T. Value reflects a balance among benefits, risk and resources, and enterprises need an actionable strategy and governance system to realize this value.	CSF states that "it can be used to help identify and prioritize actions for reducing cybersecurity risk, and it is a tool for aligning policy, business, and technological approaches to managing that risk."
2. A governance system for enterprise I&T is built from a number of components that can be of different types and that work together in a holistic way.	The CSF provides a common language for understanding, managing and expressing cybersecurity risk to internal and external stakeholders. CSF provides a common language for organizing and communicating about outcomes for EGIT cybersecurity.
3. A governance system should be dynamic. This means that each time one or more of the design factors are changed (e.g., a change in strategy or technology), the impact of these changes on the EGIT system must be considered. A dynamic view of EGIT leads toward a viable and future-proof EGIT system.	The CSF provides a flexible process by which the current, planned and future states of an enterprise may be considered in light of business drivers and risk implications. Users determine actions for EGIT based on comparison of current and desired outcomes, supporting cost-effective and forward-looking activities.
4. A governance system should clearly distinguish between governance and management activities and structures.	The CSF steps include determination of stakeholder objectives and priorities. Implementation Tiers are used to determine and achieve desired governance models through integration of a risk management strategy similar to that described in the COBIT governance objective EDM03 *Ensured risk optimization*. CSF then provides for numerous outcomes (described in the Framework Core) to achieve objectives similar to those in the four COBIT management domains.
5. A governance system should be customized to the enterprise needs, using a set of design factors as parameters to customize and prioritize the governance system components.	CSF profiles support flexibility in content and structure. Understanding the risk implications of the current state helps select cost-effective next steps. Because target-state profiles include prioritization (based on the stakeholder drivers determined in CSF Step 1), tailoring and priority setting are supported and encouraged.
6. A governance system should cover the enterprise end to end, focusing on not only the IT function but also all technology and information processing the enterprise puts in place to achieve its goals, regardless of its location in the enterprise.	CSF profiles help to collaborate among various program elements, including those of external providers. Application of CSF Functions, Categories and Subcategories supports a broad array of information and technology, helping to achieve the benefits/resource/risk balance for any systems included in the Framework scope.

2.2 Coordination of Framework Implementation

Guidance regarding stakeholder communication is central to the CSF. Industry feedback during initial CSF development indicated that risk decisions, in many enterprises, were not well aligned with enterprise drivers and goals.

Enterprises with strong governance operate with lower costs and make more efficient and effective use of resources. Further, external parties assess enterprises with strong governance as having greater internal control and lower general levels of risk.[10]

The common flow of information and decisions in the CSF at the following organizational levels are similar to those described in COBIT 2019 stakeholder roles (**figure 2.3**).

Figure 2.3—Comparison of CSF and COBIT Roles

CSF Role	COBIT 2019 Role
Executive level	Board of directors and executive management
Business/process	Business management and business process owners
Implementation/operations	IT management, IT process owners (e.g., head of operations, chief architect, IT security manager and business continuity management specialist) and other implementation team members

The executive level communicates information about enterprise goals and mission priorities, using language, approaches and communications that are meaningful to executive management. This activity is comparable to COBIT Implementation Phase 1–What Are the Drivers? Dialogue with business management and business process owners includes definition of appropriate risk tolerances and available resources. The business/process level, in turn, uses the information as inputs into the risk management process, and collaborates with IT management and IT process owners to communicate business needs.

These two levels of management determine the current cybersecurity state using a Framework Profile template (see "Framework Profiles" later in this chapter). The Current Profile and Target Profile provide considerations comparable to COBIT Implementation Phase 2–Where Are We Now? and Phase 3–Where Do We Want To Be? By comparing the target with the current state, the implementation team can recommend specific and prioritized actions to achieve stakeholder goals, aligned with Phase 1 business drivers, resource requirements and enterprise risk appetite. This action plan, comparable to COBIT Implementation Phase 4–What Needs To Be Done? and Phase 5–How Do We Get There?, provides a cost-effective, agile approach to EGIT that is scalable to any size of enterprise.

As **figure 2.4** illustrates, the information flow is cyclical, with ongoing monitoring as a critical step. COBIT Implementation Phase 6–Did We Get There? and Phase 7–How Do We Keep the Momentum Going? provide important considerations to ensure ongoing, cost-effective governance and management. For example, as technical changes occur (e.g., upgrades in physical, process and technical assets; emerging threats; newly discovered or remediated vulnerabilities), the implementation/operations level communicates the Profile implementation progress to the business/process level.

The business/process level uses this information to perform an impact assessment in consideration of business drivers. Business/process level management reports the outcomes of that impact assessment to the executive level, using language and methods appropriate to communicate with the board of directors and executive management.

[10] ISACA, *Getting Started With Governance of Enterprise IT (GEIT)*, March 2015, *http://www.isaca.org/Knowledge-Center/Research/ResearchDeliverables/Pages/getting-started-with-governance-of-enterprise-it.aspx*

Figure 2.4—Notional Information and Decision Flows Within an Organization

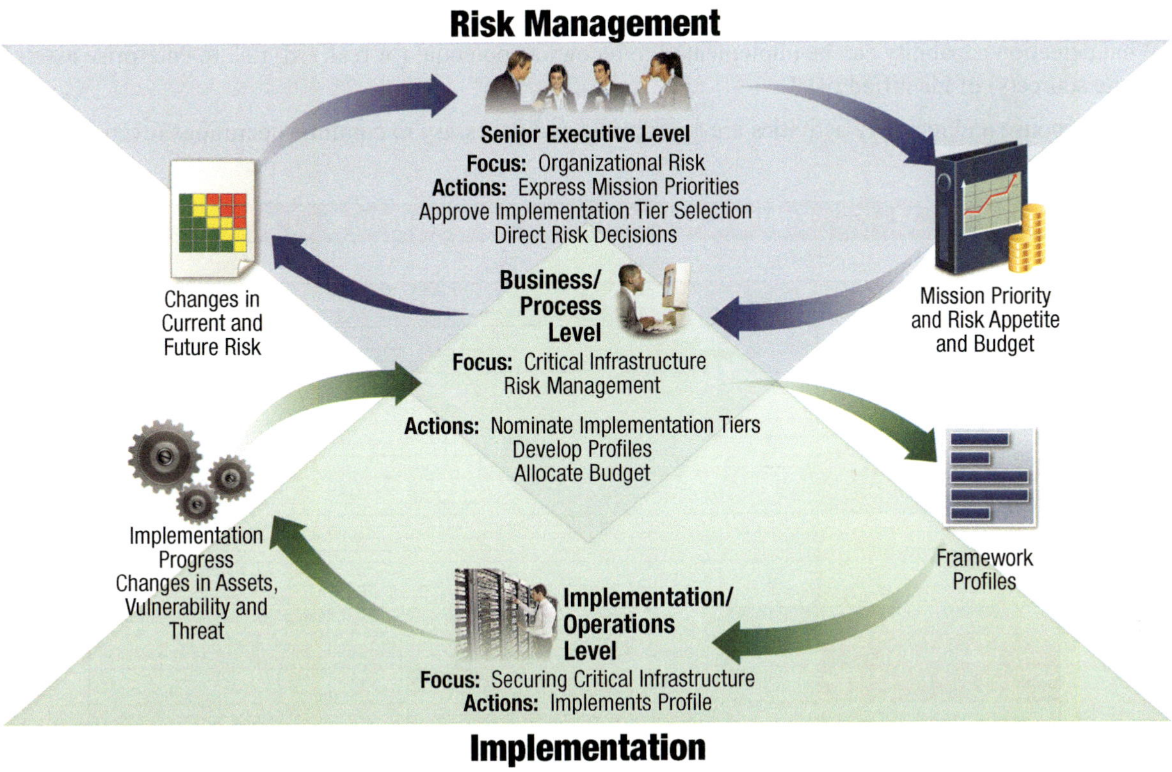

Source: NIST, *Framework for Improving Critical Infrastructure Cybersecurity*, Version 1.1, USA, 16 April 2018, *https://nvlpubs.nist.gov/nistpubs/CSWP/NIST.CSWP.04162018.pdf*

2.3 Framework Core

The Framework Core is a set of cybersecurity activities, desired outcomes and applicable references that are common across critical infrastructure sectors. The Core presents industry standards, guidelines and practices in a manner that allows for communication of cybersecurity activities and outcomes across the enterprise, from the executive level to the implementation/operations level. The Framework Core is organized around five fundamental, concurrent and continuous Functions:

- Identify (ID)
- Protect (PR)
- Detect (DE)
- Respond (RS)
- Recover (RC)

When considered together, these Functions provide a high-level, strategic view of the cybersecurity risk management life cycle of an enterprise. For each Function, the Framework Core identifies key Categories and Subcategories and maps them to representative Informative References, including standards, guidelines and practices (**figure 2.5**).

Outcomes in the Framework Core help practitioners answer the following key questions:

- What people, processes and technologies are essential to provide the right services to the right stakeholders?
- What must be done to protect those assets from risk discovered in the Identify Function?
- What detection capability can be implemented to recognize potential (or realized) risk to enterprise assets posed by the source(s) of identified risk?
- What response and recovery activities are appropriate and necessary to continue operations (even if diminished) or restore services?

Figure 2.5—Framework Core Structure

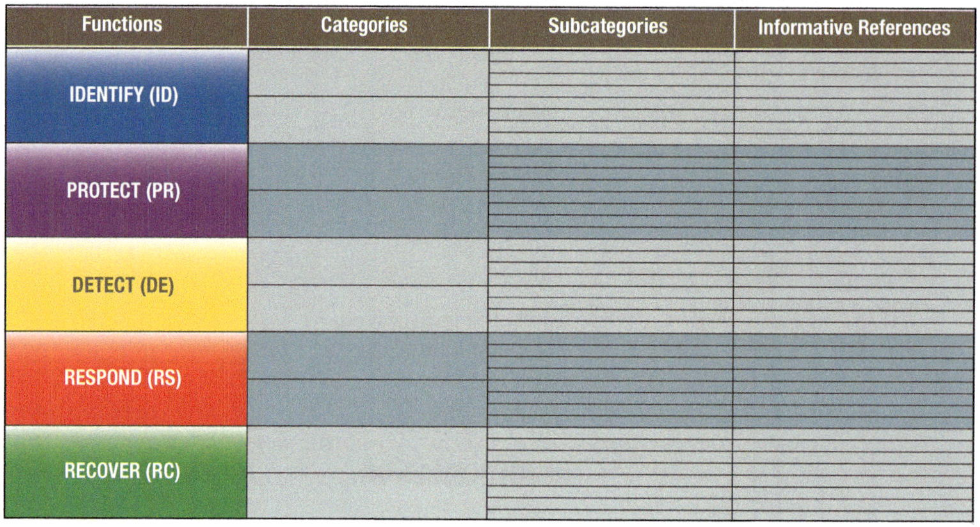

Source: NIST, *Framework for Improving Critical Infrastructure Cybersecurity*, Version 1.1, USA, 16 April 2018, *https://nvlpubs.nist.gov/nistpubs/CSWP/NIST.CSWP.04162018.pdf*

CSF describes the five Core Functions as follows:[11]

- ***Identify***—*Develop an organizational understanding to manage cybersecurity risk to systems, assets, people, data and capabilities.*

  The activities in the Identify Function are foundational for effective use of the Framework. Understanding the business context, the resources that support critical functions, and the related cybersecurity risks enables an organization to focus and prioritize its efforts, consistent with its risk management strategy and business needs. Examples of outcome Categories within this Function include: Asset Management; Business Environment; Governance; Risk Assessment; and Risk Management Strategy.

- ***Protect***—*Develop and implement appropriate safeguards to ensure delivery of critical services.*

  The Protect Function supports the ability to limit or contain the impact of a potential cybersecurity event. Examples of outcome Categories within this Function include: Identity Management and Access Control; Awareness and Training; Data Security; Information Protection Processes and Procedures; Maintenance; and Protective Technology.

- ***Detect***—*Develop and implement appropriate activities to identify the occurrence of a cybersecurity event.*

  The Detect Function enables timely discovery of cybersecurity events. Examples of outcome Categories within this Function include: Anomalies and Events; Security Continuous Monitoring; and Detection Processes.

[11] *Op cit* NIST, *Framework for Improving Critical Infrastructure Cybersecurity*, Version 1.1

- ***Respond***—*Develop and implement appropriate activities to take action regarding a detected cybersecurity incident.*

  The Respond Function supports the ability to contain the impact of a potential cybersecurity incident. Examples of outcome Categories within this Function include: Response Planning; Communications; Analysis; Mitigation; and Improvements.

- ***Recover***—*Develop and implement appropriate activities to maintain plans for resilience and to restore any capabilities or services that were impaired due to a cybersecurity incident.*

  The Recover Function supports timely recovery to normal operations to reduce the impact from a cybersecurity incident. Examples of outcome Categories within this Function include: Recovery Planning; Improvements; and Communications.

Each Function consists of one or more Categories, i.e., process-specific outcomes that support cybersecurity management (**figure 2.6**). These Categories, in turn, include numerous specific subcategories that provide a mechanism for determining the current state and target goals.

Figure 2.6—Function and Category Unique Identifiers

Function Unique Identifier	Function	Category Unique Identifier	Category
ID	Identify	ID.AM	Asset Management
		ID.BE	Business Environment
		ID.GV	Governance
		ID.RA	Risk Assessment
		ID.RM	Risk Management Strategy
		ID.SC	Supply Chain Risk Management
PR	Protect	PR.AC	Identity Management and Access Control
		PR.AT	Awareness and Training
		PR.DS	Data Security
		PR.IP	Information Protection Processes and Procedures
		PR.MA	Maintenance
		PR.PT	Protective Technology
DE	Detect	DE.AE	Anomalies and Events
		DE.CM	Security Continuous Monitoring
		DE.DP	Detection Processes
RS	Respond	RS.RP	Response Planning
		RS.CO	Communications
		RS.AN	Analysis
		RS.MI	Mitigation
		RS.IM	Improvements
RC	Recover	RC.RP	Recovery Planning
		RC.IM	Improvements
		RC.CO	Communications

Source: NIST, *Framework for Improving Critical Infrastructure Cybersecurity*, Version 1.1, USA, 16 April 2018, *https://nvlpubs.nist.gov/nistpubs/CSWP/NIST.CSWP.04162018.pdf*

In their initial responses to NIST, many enterprises reported that they already maintained internal processes and procedures to achieve the outcomes that were eventually prescribed in the Framework Core. Other enterprises requested specific guidance to implement equivalent practices. NIST recognizes that enterprises, sectors and other entities may use additional cost-effective and efficient internationally recognized guidance—including COBIT—to achieve outcomes in each NIST Function Subcategory and manage their cybersecurity risk.

These recognized guidance standards are **Informative References.** Appendix A: Framework Core maps the CSF Function Subcategories to elements in six Informative References, including COBIT. NIST has since released NIST Internal Report 8204 (NISTIR 8204)[12] which establishes an Online Informative References (OLIR) Program that expands or, in some cases, supersedes the Informative References published in CSF V1.1 Appendix A. The *Informative Reference Catalog* is located at *https://www.nist.gov/cyberframework/informative-references/informative-reference-catalog*.

2.4 Framework Implementation Tiers

The CSF includes several graduated Implementation Tiers that facilitate assessments and planning of cybersecurity activities (**figure 2.7**). The Tiers describe attributes to consider when creating a Target Profile or completing a Current Profile. CSF V1.1 added consideration of third parties and external partners when evaluating risk management characteristics. Although not considered a maturity model, the tiers nevertheless reflect a progression of increasing rigor and expertise in cybersecurity risk management:

- Tier 1: Partial
- Tier 2: Risk Informed
- Tier 3: Repeatable
- Tier 4: Adaptive

Implementation Tiers distinguish three fundamental dimensions of risk management to help enterprises reflectively model and evaluate their cybersecurity posture:

- **Risk Management Process**—Considers the level or degree to which:
  - Cybersecurity risk management practices are formalized and institutionalized
  - Prioritization of cybersecurity activities are informed by enterprise risk objectives, the threat environment and stakeholder requirements
- **Integrated Risk Management Program**—Considers cybersecurity risk awareness at the enterprise program level. Tiers distinguish degrees to which:
  - Risk-informed, management-approved processes and procedures are defined and implemented
  - Processes and procedures adapt and advance, based on information sharing and lessons learned from previous activities
- **External Participation**—Considers the level or degree to which the enterprise actively:
  - Informs partners about threat indicators, observations or events
  - Shares information with external partners to improve security before a security event occurs

[12] NIST, "Cybersecurity Framework Online Informative References (OLIR) Submissions," NISTIR 8204, April 2019, USA, *https://doi.org/10.6028/NIST.IR.8204*

Figure 2.7—Framework Implementation Tiers			
Tier	**Risk Management Process**	**Integrated Risk Management Program**	**External Participation**
Tier 1: Partial	Organizational cybersecurity risk management practices are not formalized, and risk is managed in an ad hoc and sometimes reactive manner. Prioritization of cybersecurity activities may not be directly informed by organizational risk objectives, the threat environment, or business/mission requirements.	There is limited awareness of cybersecurity risk at the organizational level. The organization implements cybersecurity risk management on an irregular, case-by-case basis due to varied experience or information gained from outside sources. The organization may not have processes that enable cybersecurity information to be shared within the organization.	The organization does not understand its role in the larger ecosystem with respect to either its dependencies or dependents. The organization does not collaborate with or receive information (e.g., threat intelligence, best practices, technologies) from other entities (e.g., buyers, suppliers, dependencies, dependents, ISAOs, researchers, governments), nor does it share information. The organization is generally unaware of the cyber supply chain risks of the products and services it provides and that it uses.
Tier 2: Risk Informed	Risk management practices are approved by management but may not be established as organizational-wide policy. Prioritization of cybersecurity activities and protection needs is directly informed by organizational risk objectives, the threat environment, or business/mission requirements.	There is an awareness of cybersecurity risk at the organizational level, but an organization-wide approach to managing cybersecurity risk has not been established. Cybersecurity information is shared within the organization on an informal basis. Consideration of cybersecurity in organizational objectives and programs may occur at some but not all levels of the organization. Cyber risk assessment of organizational and external assets occurs, but is not typically repeatable or reoccurring.	Generally, the organization understands its role in the larger ecosystem with respect to either its own dependencies or dependents, but not both. The organization collaborates with and receives some information from other entities and generates some of its own information, but may not share information with others. Additionally, the organization is aware of the cyber supply chain risks associated with the products and services it provides and uses, but does not act consistently or formally upon those risks.
Tier 3: Repeatable	The organization's risk management practices are formally approved and expressed as policy. Organizational cybersecurity practices are regularly updated based on the application of risk management processes to changes in business/mission requirements and a changing threat and technology landscape.	There is an organization-wide approach to manage cybersecurity risk. Risk-informed policies, processes, and procedures are defined, implemented as intended, and reviewed. Consistent methods are in place to respond effectively to changes in risk. Personnel possess the knowledge and skills to perform their appointed roles and responsibilities. The organization consistently and accurately monitors cybersecurity risk of organizational assets. Senior cybersecurity and non-cybersecurity executives communicate regularly regarding cybersecurity risk. Senior executives ensure consideration of cybersecurity through all lines of operation in the organization.	The organization understands its role, dependencies, and dependents in the larger ecosystem and may contribute to the community's broader understanding of risks. It collaborates with and receives information from other entities regularly that complements internally generated information, and shares information with other entities. The organization is aware of the cyber supply chain risks associated with the products and services it provides and that it uses. Additionally, it usually acts formally upon those risks, including mechanisms such as written agreements to communicate baseline requirements, governance structures (e.g., risk councils), and policy implementation and monitoring.

Figure 2.7—Framework Implementation Tiers *(cont.)*			
Tier	**Risk Management Process**	**Integrated Risk Management Program**	**External Participation**
Tier 4: Adaptive	The organization adapts its cybersecurity practices based on previous and current cybersecurity activities, including lessons learned and predictive indicators. Through a process of continuous improvement incorporating advanced cybersecurity technologies and practices, the organization actively adapts to a changing threat and technology landscape and responds in a timely and effective manner to evolving, sophisticated threats.	There is an organization-wide approach to managing cybersecurity risk that uses risk-informed policies, processes, and procedures to address potential cybersecurity events. The relationship between cybersecurity risk and organizational objectives is clearly understood and considered when making decisions. Senior executives monitor cybersecurity risk in the same context as financial risk and other organizational risks. The organizational budget is based on an understanding of the current and predicted risk environment and risk tolerance. Business units implement executive vision and analyze system-level risks in the context of the organizational risk tolerances. Cybersecurity risk management is part of the organizational culture and evolves from an awareness of previous activities and continuous awareness of activities on their systems and networks. The organization can quickly and efficiently account for changes to business/mission objectives in how risk is approached and communicated.	The organization understands its role, dependencies, and dependents in the larger ecosystem and contributes to the community's broader understanding of risks. It receives, generates, and reviews prioritized information that informs continuous analysis of its risks as the threat and technology landscapes evolve. The organization shares that information internally and externally with other collaborators. The organization uses real-time or near real-time information to understand and consistently act upon cyber supply chain risks associated with the products and services it provides and that it uses. Additionally, it communicates proactively, using formal (e.g. agreements) and informal mechanisms to develop and maintain strong supply chain relationships.

Source: NIST, *Framework for Improving Critical Infrastructure Cybersecurity*, Version 1.1, USA, 16 April 2018, *https://nvlpubs.nist.gov/nistpubs/CSWP/NIST.CSWP.04162018.pdf*

It is worth noting that the CSF provides neither descriptive guidance regarding how to measure these attributes, nor a quantitative method to determine the applicable tier. During initial CSF development, NIST received numerous comments supporting a maturity model like the Electricity Subsector Cybersecurity Capability Maturity Model (ES-C2M2).[13] However, strict criteria are difficult to apply consistently across a broad array of users, and NIST is not the authority for deciding mandatory thresholds. For those reasons, the Tiers remain subjective. Nonetheless, they help enterprises consider current risk management practices, threat environment, legal and regulatory requirements, business/mission objectives and organizational constraints. The continued omission of a concrete measurement standard in CSF version 1.1 is not intended to discourage such measurement; enterprises and industry groups (e.g., critical infrastructure sectors) may develop criteria to facilitate comparison and communication of Tier selection.

Framework Implementation Tiers are similar to COBIT 2019 capability levels. COBIT 2019 supports a CMMI-based process capability scheme.[14] Although process capability levels (PLCs) are assessed at the individual process or focus area maturity level, CSF tiers apply to the enterprise itself—or a subcomponent of the enterprise—depending

[13] US Department of Energy (DOE), *Electricity Subsector Cybersecurity Capability Maturity Model (ES-C2M2)*, February 2014, *https://www.energy.gov/sites/prod/files/2014/02/f7/ES-C2M2-v1-1-Feb2014.pdf#targetText=The%20ES%2DC2M2%20is%20part,characteristics%20of%20the%20electricity%20subsector.&targetText=Enable%20utilities%20to%20prioritize%20actions%20and%20investments%20to%20improve%20cybersecurity*

[14] *Op cit* ISACA, *COBIT® 2019 Framework: Introduction and Methodology*, pp. 38-40

on the scope of the implementation. Consideration of PCLs may assist with determination of appropriate Framework Tiers.

Rating outcomes described in **figure 2.7** require professional judgment by the implementer. The reasons for selecting a Tier, and for agreeing/disagreeing with an outcome statement in the Profiles, should be clearly documented so that advice can be given in areas where processes can be improved.

Framework Tiers and PLCs can be compared in the following ways:

- **Tier 1 (Partial)** is analogous to process capability levels 0 (Incomplete) and 1 (Initial). At these levels, risk management and information-sharing processes are either not implemented or not yet formalized to provide consistent organizational benefit.
- **Tier 2 (Risk Informed)** is analogous to process capability level 2 (Managed). Outcomes are implemented in a managed fashion, informed by organizational risk processes, and provide significant organizational awareness of cybersecurity risk management.
- **Tier 3 (Repeatable)** is analogous to process capability level 3 (Defined). The managed process is now implemented using a defined method that is capable of achieving intended outcomes.
- **Tier 4 (Adaptive)** is comparable to process capability levels 4 (Quantitative) and 5 (Optimizing). Outcomes are achieved proactively, leveraging the experience of internal and external stakeholders and/or informed through external information sources. At process capability level 5, the activities to accomplish given outcomes are continuously improved to meet relevant current and projected business goals in an optimal way.

  The role of the Tiers in determining risk approach is closely related to COBIT's governance objective EDM03 *Ensured risk optimization*. As the enterprise adapts its cybersecurity practices based on lessons learned and predictive indicators—and builds a holistic enterprise approach to risk management—it is better able to ensure identification and management of risk to enterprise value. This capability, in turn, enables EDM03 goals:

  - To ensure that I&T-related enterprise risk does not exceed risk appetite and risk tolerance
  - To identify and manage the impact of I&T risk to enterprise value
  - To minimize any potential for compliance failures

2.5 Framework Profiles

A Framework Profile represents outcomes that an enterprise has selected (based on business needs) from Core categories and subcategories. A Profile can be characterized as the alignment of standards, guidelines and practices to the Framework Core in a particular implementation scenario. Profiles can be used to identify opportunities for improving cybersecurity posture by comparing a Current Profile (current state) with a Target Profile (desired future state). To develop a Profile, an enterprise can review each of the Core categories and subcategories and, based on business drivers and a risk assessment, determine which are most important; the enterprise adds categories and subcategories as needed to address its risk. The Current Profile can then be used to support prioritization and measurement of progress toward the Target Profile, factoring in business needs including cost-effectiveness and innovation. Profiles can be used to conduct self-assessments and to communicate within an enterprise or between enterprises.

To assist enterprises in adopting and implementing the Framework, CSF section 3.2, "Establishing or Improving a Cybersecurity Program," lays out a seven-step implementation process. Each step is a logical precursor to the next, although some enterprises may conduct some steps in a different order. For example, an enterprise may adopt a Target Profile or perform a risk assessment before developing a Current Profile. These steps are summarized and aligned with detailed implementation recommendations (see chapter 3 of this publication) and should be repeated as necessary to improve an enterprise's cybersecurity.

2.6 Risk Considerations From COBIT 2019 and the CSF

Maintaining an understanding of enterprise security risk is a key component of the CSF. Step 4 of the CSF implementation process includes the requirement for performing a risk assessment. Risk assessments provide stakeholders and managers an opportunity to weigh security vulnerabilities, threats to the enterprise and technologies against operational requirements. Risk assessments help define which Core subcategories are required to mitigate risk and to identify required levels of mitigation. Different degrees of rigor in implementing cybersecurity controls can be attained by applying the Implementation Tiers described in CSF section 2.2, "Framework Implementation Tiers".

Risk assessments provide stakeholders and managers an opportunity to weigh security vulnerabilities, threats to the enterprise and technologies against operational requirements.

Whereas, ISACA defines risk as "the combination of the probability of an event and its impact," the International Organization for Standardization (ISO) defines risk as the "effect of uncertainty on objectives," noting that an effect may be positive or negative deviation from the norm.[15] Similarly, ISACA defines IT risk as business risk associated with the use, ownership, operation, involvement, influence and adoption of I&T within an enterprise. I&T risk consists of I&T-related events that could potentially impact the business.[16] IT risk can occur with uncertain frequency and impact and creates challenges in meeting strategic goals and objectives. I&T risk always exists—regardless of whether it is recognized by the enterprise. In the context of applying the CSF, then, the primary consideration is the likelihood of achieving stakeholder goals.

Managed risk enables business drivers, enhances opportunities and provides executives and managers with an understanding of the security strengths and weaknesses within the enterprise. When risk is poorly managed, business value is reduced, I&T are misused, and executives and managers remain unaware of potential security threats and vulnerabilities that can lead to lost revenue or reputation. **Figure 2.8** illustrates the inverse relation between well and poorly governed and managed I&T-related risk (on the one hand) and enterprise business value (on the other hand).

Managed risk enables business drivers, enhances opportunities and provides executives and managers with an understanding of the security strengths and weaknesses within the enterprise.

[15] ISO, Guide 73:2009, *Risk management — Vocabulary*, November 2009 (reviewed and confirmed as current in 2016), *https://www.iso.org/standard/44651.html*

[16] See *op cit* ISACA, *COBIT®2019 Framework: Introduction and Methodology*, discussion of Design Factor (DF) 3 *Risk Profile* and DF4 *I&T-Related Issues*.

Figure 2.8—Risk Duality

Source: Adapted from *COBIT® 5 for Risk*, ISACA, USA, 2013

2.7 Risk Function Perspective

COBIT 2019 regards optimization of risk as a key value objective: its governance and management must be considered as part of overall EGIT. For each governance component, risk design factors inform how the component contributes to the overall risk governance and management function:

- Processes are required to define and sustain the risk function, and govern and manage risk (e.g., EDM01 and APO01).
- Information flows are required to govern and manage risk—risk universe, risk profile, etc.
- Organizational structures are required to govern and manage risk—ERM committee, risk function, etc.

2.8 The Risk Management Perspective

The risk management perspective considers how to identify, analyze and respond to risk using COBIT 2019. This perspective requires core risk processes (EDM03 *Ensured risk optimization* and APO12 *Managed risk*) to be implemented.

The CSF is designed to reduce risk by improving the management of cybersecurity risk relative to enterprise objectives. Ideally, enterprises using the CSF will be able to measure and assign values to their risk *along with* the cost and benefits of mitigating risk to acceptable levels.[17] The CSF leverages the risk assessment process to define how enterprises will implement each Core Subcategory. Completing a risk assessment establishes the likelihood that

[17] NIST, *Framework for Improving Critical Infrastructure Cybersecurity*, Version 1.1, Section 4.0, "Self-Assessing Cybersecurity Risk with the Framework," *https://nvlpubs.nist.gov/nistpubs/CSWP/NIST.CSWP.04162018.pdf*

a risk event will occur and predicts what the resulting impact will be. Enterprises may choose to complete several risk assessments for each business area and aggregate the information to form enterprise risk assessments.

The CSF is designed to reduce risk by improving the management of cybersecurity risk relative to enterprise objectives. Ideally, enterprises using the CSF will be able to measure and assign values to their risk *along with* the cost and benefits of mitigating risk to acceptable levels.

For some enterprises, a separate risk assessment may be conducted for each business area (e.g., human resources, accounting and customer support) as defined by CSF Step 1: Prioritize and Scope. Separate risk assessments can support separate Target Profiles and ensure that risk specific to each business area is addressed appropriately, without overcompensating. The enterprisewide risk assessment defines a minimum threshold and ensures that less-sensitive business areas are not neglected and not providing avenues of attack for malicious users.

After the risk assessment is complete, enterprises can determine the acceptable level of risk for I&T assets and systems, expressed as risk tolerance. The risk tolerance is used to define controls required for each Subcategory and determine the rigor required for implementing each control.

2.9 Conclusion

Cybersecurity spending has increased by 141 percent from 2010 to 2018, yet, during the same period, the annual number of data breaches and exposed records in the United States alone has soared.[18]

The better an enterprise can measure its risk, costs and benefits of cybersecurity strategies and steps, the more rational, effective and valuable its cybersecurity approach and investments will be. Risk assessments capture a moment in time, and, therefore, must reoccur periodically for optimal utility. As cybersecurity performance metrics continue to evolve, enterprises may discover prior investments are insufficient. Clearly specifying desired business outcomes and measurements can minimize the chance of such outcomes. With regular self-assessments, enterprises can improve decision making about investment types and priorities. CSF section 4.0, "Self-Assessing Cybersecurity Risk with the Framework," provides additional information on self-assessing cybersecurity risk.

[18] For data that indicate a trend of increasing breaches and exposed records, see Clement, J.; "Cyber crime: number of breaches and records exposed 2005-2018," Statista®, 5 August 2019, *https://www.statista.com/statistics/273550/data-breaches-recorded-in-the-united-states-by-number-of-breaches-and-records-exposed/*

Chapter 3
Framework Implementation

3.1 Introduction

This chapter describes the use of ISACA methodologies to implement guidance in the CSF section 3.0, "How To Use the Framework." CSF provides seven high-level steps, and COBIT provides seven high-level phases—which generally align—although COBIT provides a post-execution assessment (Phase 6–Did We Get There?) and ongoing life cycle maintenance activities (Phase 7–How Do We Keep the Momentum Going?) that are implicit, but not fully described in the CSF. It is important to note that implementation is not an all-or-nothing proposition. Practitioners adopting the processes described should select whichever ones assist in accomplishing enterprise goals; processes are available to select from—but not all are mandatory.

Tables throughout section 3.0 describe how to apply the seven COBIT 2019 implementation phases, and how doing so helps to implement the CSF. The tables specify:

- COBIT 2019 phases
- Relevant CSF activities
- Description of the COBIT phase and its objectives
- Key tasks and activities that are relevant to COBIT and/or CSF
- Specific inputs and outputs for the phase
- Factors related to the COBIT design workflow
- Supporting practices from *COBIT® 2019 Framework: Governance and Management Objectives*

The activities and processes described are informative and may help the implementation team to determine what to do for each phase. Note that the phases are not prescriptive or necessarily obligatory, and should be tailored to achieve individual enterprise goals.

3.2 Relationship of the COBIT 2019 Goals Cascade to the CSF

The CSF recognizes that every enterprise faces unique challenges and opportunities—including having numerous internal and external stakeholders—and, thus, acknowledges that each stakeholder has unique requirements for governance and management activities. Stakeholders drive requirements for the enterprise, and, thus, for cybersecurity risk. As those requirements are set, the enterprise can use the COBIT 2019 goals cascade to formalize and refine requirements.

The COBIT 2019 goals cascade is shown in **figure 3.1.**

Figure 3.1—COBIT 2019 Goals Cascade

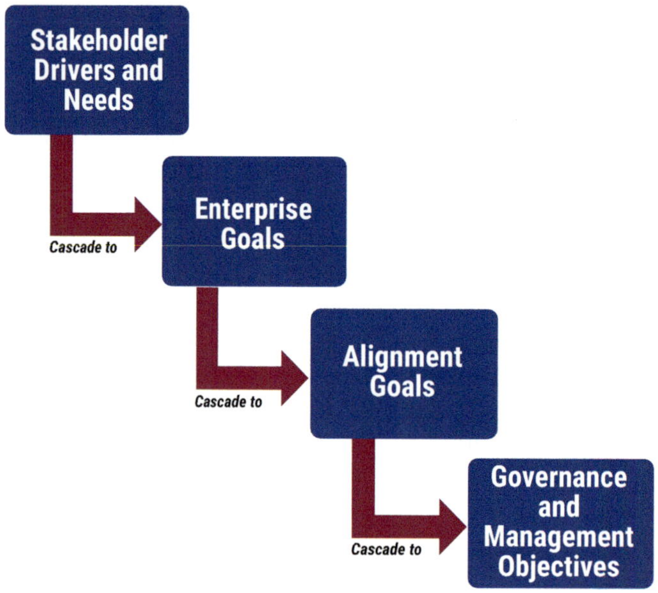

Source: ISACA, *COBIT 2019 Framework: Introduction and Methodology*, USA, 2018, figure 4.16

The goals cascade supports identification of stakeholder needs and drivers, which enhance understanding of broad enterprise goals (EGs), such as EG03 *Compliance with external laws and regulations* and EG06 *Business service continuity and availability*.[19] The achievement of enterprise goals is supported by technical outcomes and I&T-related alignment goals (AGs), which, in turn, require successful application and use of many components. The component concept is detailed in the COBIT 2019 framework. Components include processes, organizational structures and information (among others); for each component, COBIT defines a set of specific, relevant enterprise and alignment goals in support of I&T goals. In relation to the CSF, COBIT 2019 components support activities and outcomes in the Core categories and subcategories.

One important claim resonated throughout CSF development workshops: Any given enterprise may have layers of key stakeholders with varying enterprise goals. In the critical infrastructure community, for example, enterprise goals may include drivers from national priorities, stakeholders from critical sector-specific agencies or officials from sector coordinating councils. These are not unlike existing enterprise goals, such as EG03 *Compliance with external laws and regulations*. Examining the enterprise goals in this step should include understanding balanced priorities—what is best for the enterprise relative to any external commitments, such as provisioning of critical services.

3.3 COBIT Implementation Phase 1—What Are the Drivers?

The overarching purpose of this publication is to model the use of COBIT 2019 in implementing the CSF. To help orient the reader, each CSF step is reproduced in the applicable subsection discussion. The first step of the CSF follows:

[19] See *op cit* ISACA, *COBIT® 2019 Framework: Introduction and Methodology*, sections 4.6 Goals Cascade, 4.6.1 Enterprise Goals and 4.6.2 Alignment Goals.

> ***Step 1: Prioritize and Scope.*** *The organization identifies its business/mission objectives and high-level organizational priorities. With this information, the organization makes strategic decisions regarding cybersecurity implementations and determines the scope of systems and assets that support the selected business line or process. The Framework can be adapted to support the different business lines or processes within an organization, which may have different business needs and associated risk tolerance. Risk tolerances may be reflected in a target Implementation Tier.*

ISACA guidance[20] for boards of directors and executive management points out that information security governance is the responsibility of the board of directors and senior executives. Information security governance must be an integral and transparent part of enterprise governance and should align with the IT governance framework. To exercise effective enterprise and information security governance, boards and senior executives must have a clear understanding of what to expect from their enterprise's information security program. Workshops leading to the development of the CSF confirmed this guidance. Contributors to the CSF pointed out that effective alignment of business drivers with EGIT resulted in improved security and better understanding of enterprise security requirements. EGIT's principles support the use of language and terminology that are familiar to executives—rather than technical jargon and buzzwords lacking alignment with common business terms. Understanding of governance issues and benefits—conveyed in business terms—supports buy-in and commitment from senior management.

This approach helps secure Core outcomes through selected enterprise goals and processes, directly supports stakeholder goals and drivers, and transforms EGIT from a mere compliance exercise into a value engine for the enterprise. **Figure 3.2** details the implementation considerations for COBIT Implementation Phase 1 and NIST CSF Step 1.

Figure 3.2—Implementation Considerations NIST CSF Step 1
Implementation Considerations
COBIT Implementation Phase 1—What Are the Drivers? NIST CSF Step 1: Prioritize and Scope
Description
This phase identifies enterprise and alignment goals and illustrates how I&T contribute to enterprise goals via solutions and services.
Objectives
• Ensure the program team knows and understands the enterprise goals. • Identify critical processes or other components addressed in the improvement plan. • Identify the appropriate management practices for each selected process. • Obtain an understanding of the enterprise's present and future attitude toward risk. • Determine the current capability of the selected processes. • Understand the enterprise's capacity and capability for change.
Key Tasks and Activities
Recognize the need to act:
• Identify the governance context, business and I&T pain points and events. • Identify business and governance drivers. • Identify compliance requirements. • Identify priorities and business strategy dependent on I&T. • Define EGIT policy, objectives, guiding principles and high-level improvement targets.

[20] ISACA/ITGI, *Information Security Governance: Guidance for Boards of Directors and Executive Management, 2nd Edition*, USA, 2006; *http://www.isaca.org/Knowledge-Center/Research/ResearchDeliverables/Pages/Information-Security-Governance-Guidance-for-Information-Security-Managers.aspx*

Figure 3.2—Implementation Considerations NIST CSF Step 1 *(cont.)*

Implementation Considerations	
Key Tasks and Activities	
Establish the desire to change:	
• Analyze the environment in which change needs to be initiated. • Determine ongoing or planned enterprise initiatives. • Understand the breadth and depth of the change. • Identify stakeholders involved in the initiative from different areas of the enterprise. • Determine ability to implement the change.	
Initiate the program:	
• Provide high-level strategic direction and establish the high-level program. • Define and assign high-level roles within the program. • Develop an outline business case indicating success factors. • Obtain executive sponsorship.	
Inputs	**Outputs**
• Outline business case • Roles and responsibilities • Identified stakeholder map • Program wake-up call • Business and IT plans and strategies • IT process descriptions, policies, standards, procedures, technical specifications • Understanding of business and IT contributions	• Agreed alignment goals and impact • Selected governance and management objectives • Current performance levels of selected governance and management objectives • Risk acceptance position and profile • Strengths on which to build • Evaluated outline business case • Agreed understanding of the issues and challenges
COBIT 2019 Design Workflow Step 1: Understand the Enterprise Context and Strategy	
• Understand enterprise strategy • Understand enterprise goals • Understand the risk profile • Understand current I&T related issues	
Available Resources	
• *COBIT® 2019 Framework: Introduction and Methodology* (governance and management objectives, goals cascade, enterprise goals-alignment goals cascade), *www.isaca.org/cobit* • *COBIT® 2019 Framework: Governance and Management Objectives* (APO01, APO02, APO05, APO12, BAI01, BAI11, MEA01, MEA02, MEA03, MEA04—used for process selection and process capability assessment, and implementation and program planning) • *COBIT® 2019 Design Guide: Designing an Information and Technology Governance Solution* • ISACA supporting products as currently listed at *https://www.isaca.org*	

Additional Practice Guidance from COBIT 2019 *(cont.)*

COBIT 2019 Practice	Description
EDM01.01	**Evaluate the governance system.** Continually identify and engage with the enterprise's stakeholders, document an understanding of the requirements, and evaluate the current and future design of governance of enterprise I&T.
APO01 (all)	**Managed I&T Management Framework.** Design the management system for enterprise I&T based on enterprise goals and other design factors. Based on this design, implement all required components of the management system.
APO02.01	**Understand enterprise context and direction.** Understand the enterprise context (industry drivers, relevant regulations, basis for competition), its current way of working and its ambition level in terms of digitization.

Additional Practice Guidance from COBIT 2019 *(cont.)*	
COBIT 2019 Practice	**Description**
APO03.01	**Develop the enterprise architecture vision**. The architecture vision provides a first-cut, high-level description of the baseline and target architectures, covering the business, information, data, application and technology domains. The architecture vision provides the sponsor with a key tool to sell the benefits of the proposed capabilities to stakeholders within the enterprise. The architecture vision describes how the new capabilities (in line with I&T strategy and objectives) will meet enterprise goals and strategic objectives and address stakeholder concerns when implemented.
APO04.02	**Maintain an understanding of the enterprise environment.** Work with relevant stakeholders to understand their challenges. Maintain an adequate understanding of enterprise strategy, competitive environment and other constraints, so that opportunities enabled by new technologies can be identified.
APO05.01	**Determine the availability and sources of funds.** Determine potential sources of funds, different funding options and the implications of the funding source on the investment return expectations.
APO05.02	**Evaluate and select programs to fund.** Based on requirements for the overall investment portfolio mix and the I&T strategic plan and road map, evaluate and prioritize program business cases and decide on investment proposals. Allocate funds and initiate programs.
APO05.03	**Monitor, optimize and report on investment portfolio performance.** On a regular basis, monitor and optimize the performance of the investment portfolio and individual programs throughout the entire investment life cycle. Ensure continuous follow-up on the alignment of the portfolio with I&T strategy.
APO06.01	**Manage finance and accounting.** Establish and maintain a method to manage and account for all I&T-related costs, investments and depreciation as an integral part of the enterprise financial systems and accounts. Report using the enterprise's financial measurement systems.
APO06.02	**Prioritize resource allocation**. Implement a decision-making process to prioritize the allocation of resources and establish rules for discretionary investments by individual business units. Include the potential use of external service providers and consider the buy, develop and rent options.
APO06.03	**Create and maintain budgets**. Prepare a budget reflecting investment priorities based on the portfolio of I&T-enabled programs and I&T services.
APO06.04	**Model and allocate costs**. Establish and use an I&T costing model based, for example, on the service definition. This approach ensures that allocation of costs for services is identifiable, measurable and predictable, and encourages the responsible use of resources, including those provided by service providers. Regularly review and benchmark the cost/chargeback model to maintain its relevance and appropriateness for evolving business and IT activities.
APO06.05	**Manage costs.** Implement a cost management process that compares actual costs against budget. Costs should be monitored and reported. Deviations from budget should be identified in a timely manner and their impact on enterprise processes and services assessed.
APO07.01	**Acquire and maintain adequate and appropriate staffing.** Establish and maintain a method to manage and account for all I&T-related costs, investments and depreciation as an integral part of the enterprise financial systems and accounts. Report using the enterprise's financial measurement systems.
APO07.02	**Identify key IT personnel**. Identify key IT personnel. Use knowledge capture (documentation), knowledge sharing, succession planning and staff backup to minimize reliance on a single individual performing a critical job function.
APO07.03	**Maintain the skill and competencies of personnel.** Define and manage the skills and competencies required of personnel. Regularly verify that personnel have the competencies to fulfill their roles on the basis of their education, training and/or experience. Verify that these competencies are being maintained, using qualification and certification programs where appropriate. Provide employees with ongoing learning and opportunities to maintain their knowledge, skills and competencies at a level required to achieve enterprise goals.
APO08.01	**Understand business expectations**. Understand current business issues, objectives and expectations for I&T. Ensure that requirements are understood, managed and communicated, and their status agreed and approved.
APO08.03	**Manage the business relationship.** Manage the relationship between the IT service organization and its business partners. Ensure that relationship roles and responsibilities are defined and assigned, and communication is facilitated.

Additional Practice Guidance from COBIT 2019 *(cont.)*	
COBIT 2019 Practice	**Description**
APO10.01	**Identify and evaluate vendor relationships and contracts.** Continuously search for and identify vendors and categorize them into type, significance and criticality. Establish criteria to evaluate vendors and contracts. Review the overall portfolio of existing and alternative vendors and contracts.
APO10.02	**Select vendors**. Select suppliers according to a fair and formal practice to ensure a viable best fit based on specified requirements. Requirements should be optimized with input from potential suppliers.
APO11.01	**Establish a quality management system (QMS).** Establish and maintain a quality management system (QMS) that provides a standard, formal and continuous approach to quality management of information. The QMS should enable technology and business processes to align with business requirements and enterprise quality management.
APO11.02	**Focus quality management on customers.** Focus quality management on customers by determining their requirements and ensuring integration in quality management practices.
APO11.03	**Manage quality standards, practices and procedures and integrate quality management into key processes and solutions.** Identify and maintain standards, procedures and practices for key processes to guide the enterprise in meeting the intent of the agreed quality management standards (QMS). This activity should align with I&T control framework requirements. Consider certification for key processes, organizational units, products or services.
BAI01.01	**Maintain a standard approach for program management**. Maintain a standard approach for program management that enables governance and management review, decision-making and delivery-management activities. These activities should focus consistently on business value and goals (i.e., requirements, risk, costs, schedule and quality targets).
BAI01.02	**Initiate a program**. Initiate a program to confirm expected benefits and obtain authorization to proceed. This includes agreeing on program sponsorship, confirming the program mandate through approval of the conceptual business case, appointing program board or committee members, producing the program brief, reviewing and updating the business case, developing a benefits realization plan, and obtaining approval from sponsors to proceed.
BAI01.03	**Manage stakeholder engagement**. Manage stakeholder engagement to ensure an active exchange of accurate, consistent and timely information for all relevant stakeholders. This includes planning, identifying and engaging stakeholders and managing their expectations.
BAI01.04	**Develop and maintain the program plan.** Formulate a program to lay the initial groundwork. Position it for successful execution by formalizing the scope of the work and identifying deliverables that will satisfy goals and deliver value. Maintain and update the program plan and business case throughout the full economic life cycle of the program, ensuring alignment with strategic objectives and reflecting the current status and insights gained to date.
BAI01.05	**Launch and execute the program**. Launch and execute the program to acquire and direct the resources needed to accomplish the goals and benefits of the program as defined in the program plan. In accordance with stage-gate or release review criteria, prepare for stage-gate, iteration or release reviews to report progress and make the case for funding up to the following stage-gate or release review.
BAI01.07	**Manage program quality**. Prepare and execute a quality management plan, processes and practices that align with quality management standards (QMS). Describe the approach to program quality and implementation. The plan should be formally reviewed and agreed on by all parties concerned and incorporated into the integrated program plan.
BAI01.08	**Manage program risk.** Eliminate or minimize specific risk associated with programs through a systematic process of planning, identifying, analyzing, responding to, monitoring and controlling the areas or events with the potential to cause unwanted change. Define and record any risk faced by program management.

3.4 COBIT Implementation Phase 2—Where Are We Now?

Step 2 of the CSF follows:

> ***Step 2: Orient.*** *Once the scope of the cybersecurity program has been determined for the business line or process, the organization identifies related systems and assets, regulatory requirements, and overall risk approach. The organization then consults sources to identify threats and vulnerabilities applicable to those systems and assets.*

The third step in the CSF follows:

> ***Step 3: Create a Current Profile****. The organization develops a Current Profile by indicating which Category and Subcategory outcomes from the Framework Core are currently being achieved. If an outcome is partially achieved, noting this fact will help support subsequent steps by providing baseline information.*

Having identified the enterprise mission and drivers that support stakeholder objectives, the enterprise identifies related systems and assets that enable achieving those stakeholder needs. It is important to note that the CSF does not limit these systems and assets to purely IT or OT, which are subsets of the overall list of assets to be considered. Examples of assets to consider in CSF Step 2 include facilities in which technology resides, operators that ensure equipment functions safely and infrastructure that delivers products to customers. Having gained an understanding of the cascading goals—and, in particular, how the business and IT function need to deliver value from I&T in support of enterprise goals—the enterprise then identifies threats to, and vulnerabilities of, those systems and assets. This identification must be conducted with an understanding of the enterprise present and future attitude toward risk and IT risk position.

Before creating a Current Profile, practitioners should review the Framework Implementation Tiers in **figure 2.7**. Selecting the appropriate Tier that attains stakeholder needs in an optimal way determines the scale for answering the question, "Where are we now?" The goal of the process is to establish appropriate levels of governance and management to accomplish the risk objectives defined in COBIT Implementation Phase 1. Selecting an unsuitable Tier may result in a lack of sufficient processes to address risk or to coordinate with other entities. Improper selection of the highest Tier, however, may impose costly enterprisewide programs and processes whose benefits are not commensurate with defined phase 1 goals.

Conducting a dialogue to determine appropriate goals, Tiers and activities—in consideration of the unique enterprise context—is one of the key benefits of applying the CSF.

The CSF Version 1.1 Core contains 108 subcategories of outcomes, many of which are supported by one or more COBIT practices. Per the CSF, practitioners should create the Current Profile for all subcategories. Viewed through the lens of the enterprise Tier (which helps inform ***how*** an outcome should be accomplished), practitioners iterate through each Subcategory and determine the level to which that outcome has been attained to fulfill stakeholder goals. For each row in the template, determine and record the current level of achievement, as guided by the principles in COBIT 2019.

COBIT performance management (CPM) provides detailed criteria for determining how appropriate activities can achieve the required levels. In consideration of that guidance, select the appropriate level of achievement for each CSF Subcategory according to the scale in **figure 3.3**.

Figure 3.3—Performance Management Rating Scale		
Abbreviation	**Description**	**% Achieved**
N	Not achieved	0 to ≤15%
P	Partially achieved	> 15% to ≤50%
L	Largely achieved	> 50% to ≤85%
F	Fully achieved	> 85% to ≤100%
Source: Adapted from ISO 33020:2015, Section 5.3, page 6, *https://www.iso.org/standard/54195.html*, with the permission of ANSI on behalf of ISO. © ISO 2014. All rights reserved.		

Appendix B in this publication provides a full COBIT Current Profile template based on the Framework Core, including a detailed description of Current Profile data points (**figure B.2**).

Figure 3.4 details the implementation considerations for COBIT Implementation Phase 2 and NIST CSF Steps 2 and 3.

Figure 3.4—Implementation Considerations CSF Steps 2 and 3
Implementation Considerations
COBIT Implementation Phase 2—Where Are We Now? **NIST CSF Step 2: Orient** **NIST CSF Step 3: Create a Current Profile**
Description
This phase aligns I&T-related goals with enterprise strategy and risk, and prioritizes the most important goals and governance and management objectives.
Objectives
• Ensure the program team knows and understands the enterprise goals. • Identify critical processes or other components addressed in the improvement plan. • Identify the appropriate management practices for each selected process. • Obtain an understanding of the enterprise's present and future attitude toward risk. • Determine current capability of selected processes. • Understand the enterprise's capacity and capability for change.
Key Tasks and Activities
Assess current state:
• Identify key enterprise and supporting alignment goals. • Identify key governance issues and weaknesses related to the current and required future solutions and services. • Assess benefit/value enablement risk, program/project delivery and service delivery/IT operations risk. • Assess performance.
Form a powerful implementation team:
• Assemble a core team from the business and IT with the appropriate knowledge, expertise, profile, experience, credibility and authority. • Identify and manage any potential vested interests existing within the team to create the required level of trust. • Identify change agents with whom the core team can work.
Define problems and opportunities:
• Review and evaluate the outline business case, program feasibility and potential ROI. • Assign roles, responsibilities and process ownership. • Ensure commitment and support of affected stakeholders in program definition and execution. • Identify challenges and success factors.

Figure 3.4—Implementation Considerations CSF Steps 2 and 3 *(cont.)*

Implementation Considerations	
Inputs	**Outputs**
• Outline business case • Roles and responsibilities • Identified stakeholder map • Program wake-up call • Business and IT plans and strategies • IT process descriptions, policies, standards, procedures, technical specifications • Understanding of business and IT contribution	• Agreed alignment goals and impact • Selected governance and management objectives • Current performance levels of selected governance and management objectives • Risk acceptance position and profile • Strengths on which to build • Evaluated outline business case • Agreed understanding of the issues and challenges

COBIT Design Workflow Step 2: Determine the Initial Scope of the Governance System COBIT Design Workflow Step 3: Refine the Scope of the Governance System COBIT Design Workflow Step 4: Resolve Conflicts and Conclude the Governance System Design
COBIT Design Workflow Step 2 • Consider enterprise strategy. • Consider enterprise goals and apply the COBIT goals cascade. • Consider the risk profile of the enterprise. • Consider current I&T-related issues.
COBIT Design Workflow Step 3 • Consider the threat landscape. • Consider compliance requirements. • Consider the role of IT. • Consider the sourcing model. • Consider IT implementation methods. • Consider the technology adoption strategy. • Consider enterprise size.
COBIT Design Workflow Step 4 • Resolve inherent priority conflicts. • Conclude the governance system design.
Available Resources
• *COBIT® 2019 Framework: Introduction and Methodology* (governance and management objectives, goals cascade, enterprise goals-alignment goals cascade), *www.isaca.org/cobit* • *COBIT® 2019 Framework: Governance and Management Objectives* (APO01, APO02, APO05, APO12, BAI01, BAI11, MEA01, MEA02, MEA03, MEA04—used for process selection and process capability assessment, as well as implementation and program planning) • *COBIT® 2019 Design Guide: Designing an Information and Technology Governance Solution* • ISACA supporting products as currently listed at *www.isaca.org*

Additional Practice Guidance from COBIT 2019	
COBIT 2019 Practice	Description
APO02.01	**Understand enterprise context and direction**. Understand the enterprise context (industry drivers, relevant regulations, basis for competition), its current way of working and its ambition level in terms of digitization.
APO02.02	**Assess current capabilities, performance and digital maturity of the enterprise.** Assess the performance of current I&T services and develop an understanding of current business and I&T capabilities (both internal and external). Assess current digital maturity of the enterprise and its appetite for change.
APO03.02	**Define reference architecture.** The reference architecture describes the current and target architectures for the business, information, data, application and technology domains.
APO04.01	**Create an environment conducive to innovation**. Create an environment that is conducive to innovation, considering methods such as culture, reward, collaboration, technology forums, and mechanisms to promote and capture employee ideas.
APO07.02	**Identify key IT personnel**. Identify key IT personnel. Use knowledge capture (documentation), knowledge sharing, succession planning and staff backup to minimize reliance on a single individual performing a critical job function.
APO07.03	**Maintain the skills and competencies of personnel.** Define and manage the skills and competencies required of personnel. Regularly verify that personnel have the competencies to fulfill their roles on the basis of their education, training and/or experience. Verify that these competencies are being maintained, using qualification and certification programs where appropriate. Provide employees with ongoing learning and opportunities to maintain their knowledge, skills and competencies at a level required to achieve enterprise goals.
APO07.05	**Plan and track the usage of IT and business human resources.** Understand and track the current and future demand for business and IT human resources with responsibilities for enterprise I&T. Identify shortfalls and provide input into sourcing plans, enterprise and IT recruitment processes, and business and IT recruitment processes.
APO09.01	**Identify I&T services.** Analyze business requirements and the degree to which I&T-enabled services and service levels support business processes. Discuss and agree with the business on potential services and service levels. Compare potential service levels against the current service portfolio; identify new or changed services or service level options.
APO09.02	**Catalog I&T-enabled services**. Define and maintain one or more service catalogues for relevant target groups. Publish and maintain live I&T-enabled services in the service catalogs.
APO09.03	**Define and prepare service agreements.** Define and prepare service agreements based on options in the service catalogues. Include internal operational agreements.
APO11.02	**Focus quality management on customers.** Focus quality management on customers by determining their requirements and ensuring integration in quality management practices.
APO12.01	**Collect data.** Identify and collect relevant data to enable effective I&T-related risk identification, analysis and reporting.
BAI03.11	**Define IT products and services and maintain the service portfolio.** Define and agree on new or changed IT products or services and service level options. Document new or changed product and service definitions and service level options to be updated in the products and services portfolio.
BAI04.01	**Assess current availability, performance and capacity and create a baseline**. Assess availability, performance and capacity of services and resources to ensure that cost-justifiable capacity and performance are available to support business needs and deliver against service level agreements (SLAs). Create availability, performance and capacity baselines for future comparison.
BAI04.03	**Plan for new or changed service requirements**. Plan and prioritize availability, performance and capacity implications of changing business needs and service requirements.
BAI09.01	**Identify and record current assets**. Maintain an up-to-date, accurate record of all I&T assets that are required to deliver services and that are owned or controlled by the organization with an expectation of future benefit (including resources with economic value, such as hardware or software). Ensure alignment with configuration management and financial management.
BAI09.02	**Manage critical assets.** Identify assets that are critical in providing service capability. Maximize their reliability and availability to support business needs.

Additional Practice Guidance from COBIT 2019 *(cont.)*	
COBIT 2019 Practice	Description
BAI10.01	**Establish and maintain a configuration model.** Establish and maintain a logical model of the services, assets, infrastructure and recording of configuration items (CIs), including the relationships among them. Include the CIs considered necessary to manage services effectively and to provide a single, reliable description of the assets in a service.
BAI10.02	**Establish and maintain a configuration repository and baseline**. Establish and maintain a configuration management repository and create controlled configuration baselines.
BAI10.03	**Maintain and control configuration items**. Maintain an up-to-date repository of configuration items (CIs) by populating any configuration changes.
MEA03.01	**Identify external compliance requirements**. On a continuous basis, monitor changes in local and international laws, regulations and other external requirements and identify mandates for compliance from an I&T perspective.
MEA03.02	**Optimize response to external requirements.** Review and adjust policies, principles, standards, procedures and methodologies to ensure that legal, regulatory and contractual requirements are addressed and communicated. Consider adopting and adapting industry standards, codes of good practice and good practice guidance.

3.5 COBIT Implementation Phase 3—Where Do We Want to Be?

The fourth and fifth steps of the CSF follow:

> ***Step 4: Conduct a Risk Assessment***. *This assessment could be guided by the organization's overall risk management process or previous risk assessment activities. The organization analyzes the operational environment in order to discern the likelihood of a cybersecurity event and the impact that the event could have on the organization. It is important that organizations identify emerging risks and use cyber threat information from internal and external sources to gain a better understanding of the likelihood and impact of cybersecurity events.*
>
> ***Step 5: Create a Target Profile***. *The organization creates a Target Profile that focuses on the assessment of the Framework Categories and Subcategories describing the organization's desired cybersecurity outcomes. Organizations also may develop their own additional Categories and Subcategories to account for unique organizational risks. The organization may also consider influences and requirements of external stakeholders such as sector entities, customers, and business partners when creating a Target Profile. The Target Profile should appropriately reflect criteria within the target Implementation Tier.*

Based on the assessed Current Profile process capability levels and results of the goals analysis/process identification performed earlier, practitioners should determine an appropriate target capability level for each process. The chosen level should consider any relevant external and internal benchmarks (e.g., government templates or guidance). Understanding vulnerabilities and threats to assets (as determined in COBIT Implementation Phase 2), practitioners should then perform a comprehensive risk assessment to determine how best to protect the assets, detect and respond to attacks against them, and recover from any degradation or interruption.

Note that CSF Steps 3 and 5 may be reversed, depending on enterprise preference.

Viewed through the lens of the enterprise Tier (which helps inform how an outcome should be accomplished), iterate through each of the subcategories and determine the level to which that outcome should be attained to fulfill enterprise goals. For each row in the template, determine and record the target level of performance (**figure 3.5**) as guided by the principles in chapter 6, "Performance Management in COBIT," *COBIT® 2019 Framework: Introduction and Methodology*.[21]

[21] *Op cit* ISACA, *COBIT® 2019 Framework: Introduction and Methodology*

Figure 3.5—Performance Management Rating Scale		
Abbreviation	Description	% Achieved
N	Not achieved	0 to ≤15
P	Partially achieved	>15 to ≤50
L	Largely achieved	>50 to ≤85
F	Fully achieved	>85 to ≤100
Source: Adapted from ISO 33020:2015, Section 5.3, page 6, with the permission of ANSI on behalf of ISO. © ISO 2014. All rights reserved.		

Appendix B in this publication provides a full COBIT Target Profile template based on the Framework Core, including a detailed description of the Target Profile elements (**figure B.2**).

Figure 3.6 details the implementation considerations for COBIT Implementation Phase 3 and NIST CSF Steps 4 and 5.

Figure 3.6—Implementation Considerations NIST CSF Steps 4 and 5
Implementation Considerations
COBIT Implementation Phase 3—Where Do We Want to Be? **NIST CSF Step 4: Conduct a Risk Assessment** **NIST CSF Step 5: Create a Target Profile**
Description
Based on assessed current-state process capability levels, an appropriate target capability level should be determined for each process.
Objectives
• Determine the target capability for processes within governance and management objectives. • Determine gaps. • Translate gaps into improvement opportunities. • Create a detailed business case and high-level program plan from gathered information.
Key Tasks and Activities
Define target state:
• Identify and define improvement targets. • Based on performance and conformance, decide initial, ideal short- and long-term target capability levels for each process. • Analyze gaps. • Collate gaps into potential improvements. • Identify unmitigated residual risk and ensure its formal acceptance.
Describe and communicate desired outcomes:
• Secure participation. • Describe the high-level road map to achieve the vision and the involvement required of various stakeholders. • Set the tone at the top by using senior management to deliver key messages. • Use change agents to communicate informally and formally. • Capture communication feedback and adapt strategy accordingly.
Define the roadmap:
• Set program direction, scope, benefits and objectives at a high level. • Ensure alignment of objectives with business and IT strategies. • Consider risk and adjust the scope accordingly. • Consider change-enablement implications. • Obtain necessary budgets. • Define program accountabilities and responsibilities.

Figure 3.6—Implementation Considerations NIST CSF Steps 4 and 5 *(cont.)*	
Implementation Considerations	
Inputs	**Outputs**
• Agreed enterprise goals and impact on alignment goals • Current capability rating for selected processes • Risk acceptance position and risk profile • Change agents in different parts and at different levels in the enterprise • Evaluated outline business case • Internal and external capability benchmarks	• Target capability rating for selected processes • Description of improvement opportunities • Risk response document, including residual risk (not mitigated) • Change-enablement plan and objectives • Detailed business case • High-level program plan
COBIT Design Workflow Step 4: Conclude the Governance System Design	
• Resolve inherent priority conflicts. • Conclude the governance system design.	
Available Resources	
• *COBIT® 2019 Framework: Introduction and Methodology* (enterprise goals), *www.isaca.org/cobit* • *COBIT® 2019 Framework: Governance and Management Objectives* (management practices and activities for the target-state definition and gap analysis, APO01, APO02) • *COBIT® 2019 Design Guide: Designing an Information and Technology Governance Solution* • ISACA supporting products, as currently listed at *www.isaca.org*	

Additional Practice Guidance from COBIT 2019	
COBIT 2019 Practice	Description
APO02.03	**Define target digital capabilities.** Based on the understanding of enterprise context and direction, define the target I&T products and services and required capabilities. Consider reference standards, best practices and validated emerging technologies.
APO03.02	**Define reference architecture**. The reference architecture describes the current and target architectures for the business, information, data, application and technology domains.
APO04.01	**Create an environment conducive to innovation**. Create an environment that is conducive to innovation, considering methods such as culture, reward, collaboration, technology forums, and mechanisms to promote and capture employee ideas.
APO07.02	**Identify key IT personnel**. Identify key IT personnel. Use knowledge capture (documentation), knowledge sharing, succession planning and staff backup to minimize reliance on a single individual performing a critical job function.
APO07.03	**Maintain the skills and competencies of personnel.** Define and manage the skills and competencies required of personnel. Regularly verify that personnel have the competencies to fulfill their roles on the basis of their education, training and/or experience. Verify that these competencies are being maintained, using qualification and certification programs where appropriate. Provide employees with ongoing learning and opportunities to maintain their knowledge, skills and competencies at a level required to achieve enterprise goals.
APO07.05	**Plan and track the usage of IT and business human resources.** Understand and track the current and future demand for business and IT human resources with responsibilities for enterprise I&T. Identify shortfalls and provide input into sourcing plans, enterprise and IT recruitment processes, and business and IT recruitment processes.
APO09.01	**Identify I&T services**. Analyze business requirements and the degree to which I&T-enabled services and service levels support business processes. Discuss and agree with the business on potential services and service levels. Compare potential service levels against the current service portfolio; identify new or changed services or service level options.
APO09.02	**Catalog I&T-enabled services**. Define and maintain one or more service catalogues for relevant target groups. Publish and maintain live I&T-enabled services in the service catalogs.
APO09.03	**Define and prepare service agreements.** Define and prepare service agreements based on the options in the service catalogues. Include internal operational agreements.

Additional Practice Guidance from COBIT 2019 *(cont.)*	
COBIT 2019 Practice	Description
APO11.03	**Manage quality standards, practices and procedures and integrate quality management into key processes and solutions.** Identify and maintain standards, procedures and practices for key processes to guide the enterprise in meeting the intent of the agreed quality management standards (QMS). This activity should align with I&T control framework requirements. Consider certification for key processes, organizational units, products or services.
APO12.01	**Collect data.** Identify and collect relevant data to enable effective I&T-related risk identification, analysis and reporting.
BAI03.11	**Define IT products and services and maintain the service portfolio**. Define and agree on new or changed IT products or services and service level options. Document new or changed product and service definitions and service level options to be updated in the products and services portfolio.
BAI04.01	**Assess current availability, performance and capacity and create a baseline.** Assess availability, performance and capacity of services and resources to ensure that cost-justifiable capacity and performance are available to support business needs and deliver against service level agreements (SLAs). Create availability, performance and capacity baselines for future comparison.
BAI04.03	**Plan for new or changed service requirements**. Plan and prioritize availability, performance and capacity implications of changing business needs and service requirements.
BAI09.01	**Identify and record current assets**. Maintain an up-to-date, accurate record of all I&T assets that are required to deliver services and that are owned or controlled by the organization with an expectation of future benefit (including resources with economic value, such as hardware or software). Ensure alignment with configuration management and financial management.
BAI09.02	**Manage critical assets.** Identify assets that are critical in providing service capability. Maximize their reliability and availability to support business needs.
BAI10.01	**Establish and maintain a configuration model.** Establish and maintain a logical model of the services, assets, infrastructure and recording of configuration items (CIs), including the relationships among them. Include the CIs considered necessary to manage services effectively and to provide a single, reliable description of the assets in a service.
BAI10.02	**Establish and maintain a configuration repository and baseline**. Establish and maintain a configuration management repository and create controlled configuration baselines.
BAI10.03	**Maintain and control configuration items.** Maintain an up-to-date repository of configuration items (CIs) by populating any configuration changes.
DSS06 (all)	**Managed business process controls.** Define and maintain appropriate business process controls to ensure that information related to and processed by in-house or outsourced business processes satisfies all relevant information control requirements. Identify the relevant information control requirements. Manage and operate adequate input, throughput and output controls (application controls) to ensure that information and information processing satisfy these requirements.
MEA03.01	**Identify external compliance requirements.** On a continuous basis, monitor changes in local and international laws, regulations and other external requirements and identify mandates for compliance from an I&T perspective.
MEA03.02	**Optimize response to external requirements**. Review and adjust policies, principles, standards, procedures and methodologies to ensure that legal, regulatory and contractual requirements are addressed and communicated. Consider adopting and adapting industry standards, codes of good practice, and good practice guidance.

3.6 COBIT Implementation Phase 4—What Needs to Be Done?

CSF Step 6 follows:

> ***Step 6: Determine, Analyze, and Prioritize Gaps***. *The organization compares the Current Profile and the Target Profile to determine gaps. Next, it creates a prioritized action plan to address gaps—reflecting mission drivers, costs and benefits, and risks—to achieve the outcomes in the Target Profile. The organization then determines resources, including funding and workforce, necessary to address the gaps. Using Profiles in this manner encourages the organization to make informed decisions about cybersecurity activities, supports risk management, and enables the organization to perform cost-effective, targeted improvements.*

For each Subcategory in the Target Profile, consider the difference between the target level of achievement and the current level. The result of this gap assessment helps to identify enterprise strengths and weaknesses. The *COBIT® 2019 Implementation Guide* highlights several important considerations for COBIT Implementation Phase 3:

> *This phase may identify some relatively easy-to-achieve improvements such as improved training, sharing good practices and standardizing procedures. However, the gap analysis is likely to require considerable experience in business and IT-management techniques to develop practical solutions. Experience in undertaking behavioral and organizational change will also be needed.*
>
> *Understanding of process techniques, advanced business and technical expertise, and knowledge of business and system management software applications and services may be needed. To ensure that this phase is executed effectively, it is important for the team to work with the business and IT process owners and other required stakeholders, engaging internal expertise. If necessary, external advice should also be obtained. Risk that will not be mitigated after closing the gaps should be identifiable and formally accepted by management.*[22]

The opportunities for improvement should be documented in a prioritized action plan to address gaps. The plan should draw on mission drivers, a cost/benefit analysis and understanding of impact and likelihood of risk to achieve the outcomes as described in the Target Profile. The plan should also include consideration of the resources necessary to address the gaps. Using the Profile in this manner enables the enterprise to make informed decisions about cybersecurity activities, supports risk management and enables the enterprise to perform cost-effective, targeted improvements. **Figure 3.7** details the implementation considerations for COBIT Implementation Phase 4 and NIST CSF Step 6.

The plan should draw on mission drivers, a cost/benefit analysis and understanding of impact and likelihood of risk to achieve the outcomes as described in the Target Profile.

Figure 3.7—Implementation Considerations NIST CSF Step 6
Implementation Considerations
COBIT Implementation Phase 4—What Needs to Be Done? NIST CSF Step 6: Determine, Analyze and Prioritize Gaps
Description
Prioritize potential initiatives into formal and justifiable projects.
Objectives
• Translate improvement opportunities into justifiable, contributing projects. • Prioritize and focus on high-impact projects. • Integrate the improvement projects into the overall program plan. • Execute quick wins.

[22] ISACA, *COBIT® 2019 Implementation Guide: Implementing and Optimizing an Information and Technology Governance Solution*, USA, 2018, figure 6.11

Figure 3.7—Implementation Considerations NIST CSF Step 6 *(cont.)*	
Implementation Considerations	
Key Tasks and Activities	
Design and build: • Consider potential benefit and ease of implementation for each improvement. • Plot improvements onto an opportunity grid. • Focus on alternatives showing high benefit/high ease of implementation. • Consider alternatives. • Prioritize, select and analyze improvements. • Agree on projects to be included in the business case for approval. • Record unapproved projects and initiatives in a register for potential future consideration.	
Empower role players: • Obtain buy-in. • Design change-response plans. • Identify quick wins. • Build on any existing strengths identified in Phase 2 to realize quick wins, where possible. • Identify strengths in existing enterprise processes that could be leveraged.	
Develop the program plan: • Organize potential projects into the overall program. • Ensure that the program conforms to strategic goals and that I&T have a balanced set of initiatives. • Develop a change plan. • Identify and agree on metrics for measuring outcomes. • Define a portfolio of projects. • Define required deliverables. • Nominate project steering committees. • Establish project plans and reporting.	
Inputs	**Outputs**
• Target maturity rating for selected processes • Description of improvement opportunities • Risk response document • Change-enablement plan and objectives • Communication strategy and communication of the change vision • Detailed business case • Strengths identified in earlier phases	• Improvement project definitions • Defined change response plans • Identified quick wins • Record of unapproved projects • Program plan that sequences individual plans with allocated resources, priorities and deliverables • Project plans and reporting procedures enabled through committed resources such as skills and investment • Success metrics
COBIT Design Workflow (no relationship)	
Available Resources	
• *COBIT® 2019 Framework: Introduction and Methodology* (governance and management objectives, components of the governance system), *www.isaca.org/cobit* • *COBIT® 2019 Framework: Governance and Management Objectives* (APO05, APO12, BAI01, BAI11, goals and metrics) • ISACA supporting products, as currently listed at *www.isaca.org*	

Additional Practice Guidance from COBIT 2019	
COBIT 2019 Practice	Description
EDM01.02	**Direct the governance system.** Inform leaders on I&T governance principles and obtain their support, buy-in and commitment. Guide the structures, processes and practices for the governance of I&T in line with the agreed governance principles, decision-making models and authority levels. Define the information required for informed decision making.
EDM02.03	**Direct value optimization.** Direct value management principles and practices to enable optimal value realization from I&T-enabled investments throughout their full economic life cycle.
EDM03.02	**Direct risk management**. Direct the establishment of risk management practices to provide reasonable assurance that I&T risk management practices are appropriate and that actual I&T risk does not exceed the board's risk appetite.
EDM04.02	**Direct resource management**. Ensure the adoption of resource management principles to enable optimal use of business and I&T resources throughout their full economic life cycle.
EDM05.02	**Direct stakeholder engagement, communication and reporting**. Ensure the establishment of effective stakeholder involvement, communication and reporting, including mechanisms for ensuring the quality and completeness of information, overseeing mandatory reporting, and creating a communication strategy for stakeholders.
APO02.05	**Define the strategic plan and road map.** Develop a holistic digital strategy, in cooperation with relevant stakeholders, and detail a road map that defines the incremental steps required to achieve the goals and objectives. Ensure focus on the transformation journey through the appointment of a person who helps spearhead the digital transformation and drives alignment between business and I&T.
APO02.06	**Communicate the I&T strategy and direction.** Create awareness and understanding of the business and I&T objectives and direction, as captured in the I&T strategy, through communication to appropriate stakeholders and users throughout the enterprise.
APO08.04	**Coordinate and communicate**. Work with all relevant stakeholders and coordinate the end-to-end delivery of I&T services and solutions provided to the business.
APO11.05	**Maintain continuous improvement.** Maintain and regularly communicate an overall quality plan that promotes conintious improvement. The plan should define the need for, and benefits of, continuous improvement. Collect and analyze data about the quality management system (QMS) and improve its effectiveness. Correct nonconformities to prevent recurrence.
BAI02.04	**Obtain approval of requirements and solutions.** Coordinate feedback from affected stakeholders. At predetermined key stages, obtain approval and sign-off from the business sponsor or product owner regarding functional and technical requirements, feasibility studies, risk analyses and recommended solutions.
BAI03.01	**Design high-level solutions.** Develop and document high-level designs for the solution in terms of technology, business processes and workflows. Use agreed and appropriate phased or rapid Agile development techniques. Ensure alignment with the I&T strategy and enterprise architecture. Reassess and update the designs when significant issues occur during detailed design or building phases, or as the solution evolves. Apply a user-centric approach; ensure that stakeholders actively participate in the design and approve each version.
BAI03.02	**Design detailed solution components**. Develop, document and elaborate detailed designs progressively. Use agreed and appropriate phased or rapid Agile development techniques, addressing all components (business processes and related automated and manual controls, supporting I&T applications, infrastructure services and technology products, and partners/suppliers). Ensure that the detailed design includes internal and external service level agreements (SLAs) and operational level agreements (OLAs).
BAI03.03	**Develop solution components**. Develop solution components progressively in a separate environment, in accordance with detailed designs following standards and requirements for development and documentation, quality assurance (QA), and approval. Ensure that all control requirements in the business processes, supporting I&T applications and infrastructure services, services and technology products, and partner/vendor services are addressed.
BAI03.04	**Procure solution components**. Procure solution components, based on the acquisition plan, in accordance with requirements and detailed designs, architecture principles and standards, and the enterprise's overall procurement and contract procedures, QA requirements, and approval standards. Ensure that all legal and contractual requirements are identified and addressed by the vendor.

Additional Practice Guidance from COBIT 2019 *(cont.)*	
COBIT 2019 Practice	Description
BAI03.05	**Build solutions.** Install and configure solutions and integrate with business process activities. During configuration and integration of hardware and infrastructure software, implement control, security, privacy and auditability measures to protect resources and ensure availability and data integrity. Update the product or services catalogue to reflect the new solutions.
BAI03.06	**Perform quality assurance (QA).** Develop, resource and execute a QA plan aligned with the QMS to obtain the quality specified in the requirements definition and the enterprise's quality policies and procedures.
BAI03.07	**Prepare for solution testing.** Establish a test plan and required environments to test the individual and integrated solution components. Include the business processes and supporting services, applications and infrastructure.
BAI03.08	**Execute solution testing.** During development, execute testing continually (including control testing), in accordance with the defined test plan and development practices in the appropriate environment. Engage business process owners and end users in the test team. Identify, log and prioritize errors and issues identified during testing.
BAI05.01	**Establish the desire to change**. Understand the scope and impact of the desired change. Assess stakeholder readiness and willingness to change. Identify actions that will motivate stakeholder acceptance and participation to make the change work successfully.
BAI05.02	**Form an effective implementation team**. Establish an effective implementation team by assembling appropriate members, creating trust, and establishing common goals and effectiveness measures.
BAI05.03	**Communicate desired vision.** Communicate the desired vision for the change in the language of those affected by it. The communication should be made by senior management and include the rationale for, and benefits of, the change; the impacts of not making the change; and the vision, the road map and the involvement required of the various stakeholders.
BAI05.04	**Empower role players and identify short-term wins.** Empower those with implementation roles by assigning accountability. Provide training and align organizational structures and HR processes. Identify and communicate short-term wins that are important from a change-enablement perspective.
BAI05.05	**Enable operation and use.** Plan and implement all technical, operational and usage aspects so all those who are involved in the future state environment can exercise their responsibility.
BAI05.06	**Embed new approaches.** Embed new approaches by tracking implemented changes, assessing the effectiveness of the operation and use plan, and sustaining ongoing awareness through regular communication. Take corrective measures as appropriate (which may include enforcing compliance).
MEA01.01	**Establish a monitoring approach.** Engage with stakeholders to establish and maintain a monitoring approach to define the objectives, scope and method for measuring business solution and service delivery and contribution to enterprise objectives. Integrate this approach with the corporate performance management system.
MEA01.02	**Set performance and conformance targets**. Work with stakeholders to define, periodically review, update and approve performance and conformance targets within the performance measurement system.

3.7 COBIT Implementation Phase 5—How Do We Get There?

This step/phase commences execution of the prioritized action plan, as defined in the previous step/phase.

CSF Step 7 follows:

> ***Step 7: Implement Action Plan***. *The organization determines which actions to take to address the gaps, if any, identified in the previous step and then adjusts its current cybersecurity practices in order to achieve the Target Profile. For further guidance, the Framework identifies example Informative References regarding the Categories and Subcategories, but organizations should determine which standards, guidelines, and practices, including those that are sector specific, work best for their needs.*

Action plan execution provides an opportunity for frequent stakeholder communication, which should use language and terminology appropriate for each audience. For example, IT management discussions may consider specific facilities and processes, while board and executive discussions may be more related to annualized loss expectancy or market opportunities.

Action plan execution may be gradually implemented, building on the momentum of project success, building further credibility and improving success. Execution of the action plan provides an opportunity to foster an effective risk management culture throughout the enterprise. Performance measures and incremental metrics help with documenting success and supporting any adjustments required. Many such measures are described in COBIT 2019 processes, especially those in the Build, Acquire and Implement (BAI) and Deliver, Service and Support (DSS) domains. **Figure 3.8** details the implementation considerations for COBIT Implementation Phase 5 and NIST CSF Step 7.

Figure 3.8—Implementation Considerations NIST CSF Step 7
Implementation Considerations
COBIT Implementation Phase 5—How Do We Get There? NIST CSF Step 7: Implement Action Plan
Description
The approved improvement projects are now ready for implementation. Solutions defined by the program can now be acquired or developed and implemented into the enterprise.
Objectives
• Implement the detailed improvement projects. • Leverage enterprise program and project management capabilities, standards and practices. • Monitor, measure and report on project progress.
Key Tasks and Activities
Implement improvements:
• Develop and/or acquire solutions that include the fullscope of activities required. • Adopt and adapt available guidance to suit the enterprise's approach to policies and procedures. • Test the practicality and suitability of the solutions in the real working environment. • Roll out the solutions, considering any existing processes and migration requirements.
Enable operation and use:
• Build on the momentum and credibility. • Communicate quick-win successes and recognize and reward those involved. • Implement the change response plans. • Communicate roles and responsibilities. • Define measures of success. • Close the loop and ensure that all change requirements have been addressed. • Monitor the change enablement effectiveness and take corrective action where necessary.
Execute the plan:
• Ensure that execution of the program is based on an up-to-date and integrated (business and IT) plan of projects within the program. • Direct and monitor the contributions of all the projects in the program. • Provide regular update reports to stakeholders. • Document and monitor significant program risk and issues and agree on remediation actions. • Approve any major changes to the program and project plans.

Figure 3.8—Implementation Considerations NIST CSF Step 7 *(cont.)*	
Implementation Considerations	
Inputs	**Outputs**
• Improvement project definitions • Defined change response plans • Identified quick wins • Record of unapproved projects • Program plan with allocated resources, priorities and deliverables • Project plans and reporting procedures • Success metrics • Project definitions, plans, change strategy and response plans • Integrated program and project plans	• Implemented improvements • Implemented change response plans • Realized quick wins and visibility of change success • Success communications • Defined and communicated roles and responsibilities in the business-as-usual (BAU) environment • Project change logs and issue/risk logs • Defined business success measures • Benefits tracked to monitor realization
COBIT Design Workflow (no relationship)	
Available Resources	
• *COBIT® 2019 Framework: Governance and Management Objectives* (all objectives as good practice input; BAI01, BAI11), *www.isaca.org/cobit* • ISACA supporting products, as currently listed at *www.isaca.org*	

Additional Practice Guidance from COBIT 2019	
COBIT 2019 Practice	**Description**
EDM01.02	**Direct the governance system.** Inform leaders on I&T governance principles and obtain their support, buy-in and commitment. Guide the structures, processes and practices for the governance of I&T in line with the agreed governance principles, decision-making models and authority levels. Define the information required for informed decision making.
EDM02.03	**Direct value optimization.** Direct value management principles and practices to enable optimal value realization from I&T-enabled investments throughout their full economic life cycle.
EDM03.02	**Direct risk management.** Direct the establishment of risk management practices to provide reasonable assurance that I&T risk management practices are appropriate and that actual I&T risk does not exceed the board's risk appetite.
EDM04.02	**Direct resource management**. Ensure the adoption of resource management principles to enable optimal use of business and I&T resources throughout their full economic life cycle.
EDM05.02	**Direct stakeholder engagement, communication and reporting**. Ensure the establishment of effective stakeholder involvement, communication and reporting, including mechanisms for ensuring the quality and completeness of information, overseeing mandatory reporting, and creating a communication strategy for stakeholders.
APO02.05	**Define the strategic plan and road map.** Develop a holistic digital strategy, in cooperation with relevant stakeholders, and detail a road map that defines the incremental steps required to achieve the goals and objectives. Ensure focus on the transformation journey through the appointment of a person who helps spearhead the digital transformation and drives alignment between business and I&T.
APO02.06	**Communicate the I&T strategy and direction.** Create awareness and understanding of the business and I&T objectives and direction, as captured in the I&T strategy, through communication to appropriate stakeholders and users throughout the enterprise.
APO08.04	**Coordinate and communicate.** Work with all relevant stakeholders and coordinate the end-to-end delivery of I&T services and solutions provided to the business.
APO11.05	**Maintain continuous improvement.** Maintain and regularly communicate an overall quality plan that promotes continuous improvement. The plan should define the need for, and benefits of, continuous improvement. Collect and analyze data about the quality management system (QMS) and improve its effectiveness. Correct nonconformities to prevent recurrence.

Additional Practice Guidance from COBIT 2019 *(cont.)*	
COBIT 2019 Practice	**Description**
BAI02.04	**Obtain approval of requirements and solutions**. Coordinate feedback from affected stakeholders. At predetermined key stages, obtain approval and sign-off from the business sponsor or product owner regarding functional and technical requirements, feasibility studies, risk analyses and recommended solutions.
BAI03.01	**Design high-level solutions.** Develop and document high-level designs for the solution in terms of technology, business processes and workflows. Use agreed and appropriate phased or rapid Agile development techniques. Ensure alignment with the I&T strategy and enterprise architecture. Reassess and update the designs when significant issues occur during detailed design or building phases, or as the solution evolves. Apply a user-centric approach; ensure that stakeholders actively participate in the design and approve each version.
BAI03.02	**Design detailed solution components.** Develop, document and elaborate detailed designs progressively. Use agreed and appropriate phased or rapid Agile development techniques, addressing all components (business processes and related automated and manual controls, supporting I&T applications, infrastructure services and technology products, and partners/suppliers). Ensure that the detailed design includes internal and external service level agreements (SLAs) and operational level agreements (OLAs).
BAI03.03	**Develop solution components.** Develop solution components progressively in a separate environment, in accordance with detailed designs following standards and requirements for development and documentation, quality assurance (QA), and approval. Ensure that all control requirements in the business processes, supporting I&T applications and infrastructure services, services and technology products, and partner/vendor services are addressed.
BAI03.04	**Procure solution components**. Procure solution components, based on the acquisition plan, in accordance with requirements and detailed designs, architecture principles and standards, and the enterprise's overall procurement and contract procedures, QA requirements, and approval standards. Ensure that all legal and contractual requirements are identified and addressed by the vendor.
BAI03.05	**Build solutions.** Install and configure solutions and integrate with business process activities. During configuration and integration of hardware and infrastructure software, implement control, security, privacy and auditability measures to protect resources and ensure availability and data integrity. Update the product or services catalogue to reflect the new solutions.
BAI03.06	**Perform quality assurance (QA).** Develop, resource and execute a QA plan aligned with the QMS to obtain the quality specified in the requirements definition and the enterprise's quality policies and procedures.
BAI03.07	**Prepare for solution testing.** Establish a test plan and required environments to test the individual and integrated solution components. Include the business processes and supporting services, applications and infrastructure.
BAI03.08	**Execute solution testing.** During development, execute testing continually (including control testing), in accordance with the defined test plan and development practices in the appropriate environment. Engage business process owners and end users in the test team. Identify, log and prioritize errors and issues identified during testing.
BAI05.01	**Establish the desire to change.** Understand the scope and impact of the desired change. Assess stakeholder readiness and willingness to change. Identify actions that will motivate stakeholder acceptance and participation to make the change work successfully.
BAI05.02	**Form an effective implementation team**. Establish an effective implementation team by assembling appropriate members, creating trust, and establishing common goals and effectiveness measures.
BAI05.03	**Communicate desired vision.** Communicate the desired vision for the change in the language of those affected by it. The communication should be made by senior management and include the rationale for, and benefits of, the change; the impacts of not making the change; and the vision, the road map and the involvement required of the various stakeholders.
BAI05.04	**Empower role players and identify short-term wins.** Empower those with implementation roles by assigning accountability. Provide training and align organizational structures and HR processes. Identify and communicate short-term wins that are important from a change-enablement perspective.
BAI05.05	**Enable operation and use.** Plan and implement all technical, operational and usage aspects so all those who are involved in the future state environment can exercise their responsibility.

Additional Practice Guidance from COBIT 2019 *(cont.)*	
COBIT 2019 Practice	**Description**
BAI05.06	**Embed new approaches**. Embed new approaches by tracking implemented changes, assessing the effectiveness of the operation and use plan, and sustaining ongoing awareness through regular communication. Take corrective measures as appropriate (which may include enforcing compliance.)
MEA01.01	**Establish a monitoring approach.** Engage with stakeholders to establish and maintain a monitoring approach to define the objectives, scope and method for measuring business solution and service delivery and contribution to enterprise objectives. Integrate this approach with the corporate performance management system.
MEA01.02	**Set performance and conformance targets**. Work with stakeholders to define, periodically review, update and approve performance and conformance targets within the performance measurement system.
MEA01.03	**Collect and process performance and conformance data.** Collect and process timely and accurate data aligned with enterprise approaches.
DSS01.01	**Perform operational procedures**. Maintain and perform operational procedures and operational tasks reliably and consistently.
DSS01.02	**Manage outsourced I&T services.** Manage the operation of outsourced I&T services to maintain the protection of enterprise information and reliability of service delivery.
DSS01.04	**Manage the environment**. Maintain measures for protection against environmental factors. Install specialized equipment and devices to monitor and control the environment.
DSS01.05	**Manage facilities.** Manage facilities, including power and communications equipment, in line with laws and regulations, technical and business requirements, vendor specifications, and health and safety guidelines.
DSS02.02	**Record, classify and prioritize requests and incidents.** Identify, record and classify service requests and incidents and assign a priority according to business criticality and service agreements.
DSS02.03	**Verify, approve and fulfill service requests**. Select the appropriate request procedures and verify that the service requests fulfill defined request criteria. Obtain approval, if required, and fulfill the requests.
DSS02.04	**Investigate, diagnose and allocate incidents**. Identify and record incident symptoms, determine possible causes, and allocate for resolution.
DSS02.05	**Resolve and recover from incidents**. Document, apply and test the identified solutions or workarounds. Perform recovery actions to restore the I&T-related service.
DSS02.06	**Close service requests and incidents**. Verify satisfactory incident resolution and/or fulfillment of requests, and close.
DSS02.07	**Track status and produce reports.** Regularly track, analyze and report incidents and fulfillment of requests. Examine trends to provide information for continual improvement.
DSS03.01	**Identify and classify problems**. Define and implement criteria and procedures to identify and report problems. Include problem classification, categorization and prioritization.
DSS03.02	**Investigate and diagnose problems.** Investigate and diagnose problems using relevant subject matter experts to assess and analyze root causes.
DSS03.03	**Raise known errors.** As soon as root causes of problems are identified, create known-error records, document appropriate workarounds and identify potential solutions.
DSS03.04	**Resolve and close problems**. Identify and initiate sustainable solutions addressing the root cause. Raise change requests via the established change management process, if required, to resolve errors. Ensure that the personnel affected are aware of the actions taken and the plans developed to prevent future incidents from occurring.
DSS03.05	**Perform proactive problem management**. Collect and analyze operational data (especially incident and change records) to identify emerging trends that may indicate problems. Log problem records to enable assessment.
DSS04.02	**Maintain business resilience**. Evaluate business resilience options and choose a cost-effective and viable strategy that will ensure enterprise continuity, disaster recovery and incident response in the face of a disaster or other major incident or disruption.
DSS04.03	**Develop and implement a business continuity response.** Develop a business continuity plan (BCP) and disaster recovery plan (DRP) based on the strategy. Document all procedures necessary for the enterprise to continue critical activities in the event of an incident.

Additional Practice Guidance from COBIT 2019 *(cont.)*	
COBIT 2019 Practice	**Description**
DSS04.04	**Exercise, test and review the business continuity plan (BCP) and disaster response plan (DRP).** Test continuity on a regular basis to exercise plans against predetermined outcomes, uphold business resilience and allow innovative solutions to be developed.
DSS04.05	**Review, maintain and improve the continuity plans.** Conduct a management review of the continuity capability at regular intervals to ensure its continued suitability, adequacy and effectiveness. Manage changes to the plans in accordance with the change control process to ensure that continuity plans are kept up to date and continually reflect actual business requirements.
DSS04.06	**Conduct continuity plan training**. Provide all concerned internal and external parties with regular training sessions regarding procedures and their roles and responsibilities in case of disruption.
DSS04.07	**Manage backup arrangements.** Maintain availability of business-critical information.
DSS04.08	**Conduct post-resumption review.** Assess the adequacy of the business continuity plan (BCP) and disaster response plan (DRP) following successful resumption of business processes and services after a disruption.
DSS05.01	**Protect against malicious software**. Implement and maintain preventive, detective and corrective measures (especially up-to-date security patches and virus control) across the enterprise to protect information systems and technology from malicious software (e.g., ransomware, malware,viruses, worms, spyware, spam).
DSS05.02	**Manage network and connectivity security**. Use security measures and related management procedures to protect information over all methods of connectivity.
DSS05.03	**Manage endpoint security.** Ensure that endpoints (e.g., laptop, desktop, server, and other mobile and network devices or software) are secured at a level that is equal to or greater than the defined security requirements for the information processed, stored or transmitted.
DSS05.04	**Manage user identity and logical access.** Ensure that all users have information access rights in accordance with business requirements. Coordinate with business units that manage their own access rights within business processes.
DSS05.05	**Manage physical access to I&T assets.** Define and implement procedures (including emergency procedures) to grant, limit and revoke access to premises, buildings and areas, according to business need. Access to premises, buildings and areas should be justified, authorized, logged and monitored. This requirement applies to all persons entering the premises, including staff, temporary staff, clients, vendors, visitors or any other third party.
DSS05.06	**Manage sensitive documents and output devices**. Establish appropriate physical safeguards, accounting practices and inventory management regarding sensitive I&T assets, such as special forms, negotiable instruments, special-purpose printers or security tokens.
DSS05.07	**Manage vulnerabilities and monitor the infrastructure for security-related events.** Using a portfolio of tools and technologies (e.g., intrusion detection tools), manage vulnerabilities and monitor the infrastructure for unauthorized access. Ensure that security tools, technologies and detection are integrated with general event monitoring and incident management.
DSS06.02	**Control the processing of information.** Operate the execution of the business process activities and related controls, based on enterprise risk. Ensure that information processing is valid, complete, accurate, timely and secure (i.e., reflects legitimate and authorized business use).
DSS06.03	**Manage roles, responsibilities, access privileges and levels of authority**. Manage business roles, responsibilities, levels of authority and segregation of duties needed to support the business process objectives. Authorize access to all information assets related to business information processes, including those under the custody of the business, IT and third parties. This ensures that the business knows where the data are and who is handling data on its behalf.
DSS06.04	**Manage errors and exceptions.** Manage business process exceptions and errors and facilitate remediation, executing defined corrective actions and escalating as necessary. This treatment of exceptions and errors provides assurance of the accuracy and integrity of the business information process.
DSS06.05	**Ensure traceability and accountability for information events**. Ensure that business information can be traced to an originating business event and associated with accountable parties. This discoverability provides assurance that business information is reliable and has been processed in accordance with defined objectives.

Additional Practice Guidance from COBIT 2019 *(cont.)*	
COBIT 2019 Practice	Description
DSS06.06	**Secure information assets.** Secure information assets accessible by the business through approved methods, including information in electronic form (e.g., portable media devices, user applications and storage devices, or other methods that create new assets in any form), information in physical form (e.g., source documents or output reports) and information during transit. This benefits the business by providing end-to-end safeguarding of information.
MEA01.05	**Ensure the implementation of corrective actions**. Assist stakeholders in identifying, initiating and tracking corrective actions to address anomalies.
MEA02.02	**Review effectiveness of business process controls.** Review the operation of controls, including monitoring and test evidence, to ensure that controls within business processes operate effectively. Include activities to maintain evidence of the effective operation of controls through mechanisms such as periodic testing, continuous monitoring, independent assessments, command and control centers, and network operation centers. This evidence assures the enterprise that controls meet requirements related to business, regulatory and social responsibilities.
MEA02.03	**Perform control self-assessments.** Encourage management and process owners to improve controls proactively through a continuing program of self-assessment that evaluates the completeness and effectiveness of management's control over processes, policies and contracts.
MEA02.04	**Identify and report control deficiencies.** Identify control deficiencies and analyze and identify their underlying root causes. Escalate control deficiencies and report to stakeholders.
MEA03.03	**Confirm external compliance**. Confirm compliance of policies, principles, standards, procedures and methodologies with legal, regulatory and contractual requirements.
MEA03.04	**Obtain assurance of external compliance**. Obtain and report assurance of compliance and adherence with policies, principles, standards, procedures and methodologies. Confirm that corrective actions to address compliance gaps are closed in a timely manner.

3.8 COBIT Implementation Phase 6—Did We Get There?

COBIT 2019 Phase 6 provides mechanisms to review the execution of the action plan and consider performance regarding the monitoring approach previously established (e.g., MEA01 processes from phases 4 and 5). Implementers should consider how well the organization achieved performance and conformance targets and should update ongoing improvement and communication activities in accordance with established change management processes. This review phase provides the opportunity to share both positive and negative results with stakeholders, fostering confidence in planned solutions and ensuring alignment with organizational drivers and goals.

Implementers should consider how well the organization achieved performance and conformance targets and should update ongoing improvement and communication activities in accordance with established change management processes.

Performance and conformance data may be shared with internal teams to improve subsequent processes. Appropriately sanitized risk, activity and performance results may be shared with external partners (consistent with the organizations' document classification policy for public documents), to help improve general understanding of I&T risk management. **Figure 3.9** details the implementation considerations for COBIT Implementation Phase 6 and the NIST CSF Lifecycle Action Plan Review.

Figure 3.9—Implementation Considerations NIST CSF Lifecycle Action Plan Review

Implementation Considerations	
COBIT Implementation Phase 6—Did We Get There? NIST CSF Lifecycle Action Plan Review (not a CSF step)	
Description	
Monitor the described program improvements via alignment goals and process goals, using suitable techniques such as an IT balanced scorecard (BSC) and benefits register to verify that outcomes have been achieved.	
Objectives	
• Integrate the metrics for project performance and benefits realization.	
Key Tasks and Activities	
Operate and measure:	
• Set targets for each metric. • Obtain and gather measures. • Investigate variances. • Develop and agree on proposed corrective measures. • Adjust long-term targets based on experience. • Communicate results from performance monitoring to all interested stakeholders.	
Embed new approaches:	
• Instill new ways of working into the enterprise's culture. • Leverage pockets of excellence to provide a source of inspiration. • Maintain the communication strategy to achieve ongoing awareness and highlight successes. • Ensure open communication among all participants and stakeholders to resolve issues.	
Realize benefits:	
• Monitor performance against objectives. • Monitor investment performance. • Document lessons learned.	
Inputs	**Outputs**
• Implemented improvements • Implemented change response plans • Realized quick wins and success communications • Defined and communicated roles and responsibilities • Project change and issue/risk logs • Defined business and perception success measures • Alignment goals and IT process goals • Existing measures and/or scorecards • Business case benefits • Change response plans and strategy	• Updated project and program scorecards • Change effectiveness measures (both business and perception measures) • Report explaining scorecard results • Improvements entrenched in operations • Key metrics added into ongoing IT performance measurement approach
COBIT Design Workflow (no relationship)	
Available Resources	
• *COBIT® 2019 Framework: Governance and Management Objectives* (all objectives as good practice input, and EDM05, APO05, BAI01, BAI11, MEA01), *www.isaca.org/cobit* • ISACA supporting products, as currently listed at *www.isaca.org*	

Additional Guidance Practice from COBIT 2019	
COBIT 2019 Practices	Description
APO02.02	**Assess current capabilities, performance and digital maturity of the enterprise.** Assess the performance of current I&T services and develop an understanding of current business and I&T capabilities (both internal and external). Assess current digital maturity of the enterprise and its appetite for change.
MEA01.05	**Ensure the implementation of corrective actions.** Assist stakeholders in identifying, initiating and tracking corrective actions to address anomalies.
MEA02.02	**Review effectiveness of business process controls.** Review the operation of controls, including monitoring and test evidence, to ensure that controls within business processes operate effectively. Include activities to maintain evidence of the effective operation of controls through mechanisms such as periodic testing, continuous monitoring, independent assessments, command and control centers, and network operation centers. This evidence assures the enterprise that controls meet requirements related to business, regulatory and social responsibilities.
MEA02.03	**Perform control self-assessments**. Encourage management and process owners to improve controls proactively through a continuing program of self-assessment that evaluates the completeness and effectiveness of management's control over processes, policies and contracts.
MEA02.04	**Identify and report control deficiencies**. Identify control deficiencies and analyze and identify their underlying root causes. Escalate control deficiencies and report to stakeholders.
MEA03.04	**Obtain assurance of external compliance**. Obtain and report assurance of compliance and adherence with policies, principles, standards, procedures and methodologies. Confirm that corrective actions to address compliance gaps are closed in a timely manner.

3.9 COBIT Implementation Phase 7—How Do We Keep the Momentum Going?

An effective framework for EGIT addresses the complete life cycle of IT investment, ensuring that it creates value in alignment with enterprise objectives. Combining CSF principles and COBIT 2019 practices helps to ensure value, manage risk and support mission drivers in accordance with direction and support from the board of directors and executive management.

COBIT Implementation Phase 7 provides the opportunity to close the loop for communication workflow, as introduced in chapter 2 of this publication. As technical assessment is reported (e.g., through performance metrics such as those established through process MEA01) to business process owners, they, in turn, report progress toward enterprise goals and mission priorities, using language, approaches and communications that are meaningful to executive management. Momentum, gained by progress and effective communication, drives subsequent iterations of the life cycle. Updated challenges and opportunities lead to updated risk assessments and priorities, fostering enterprise commitment and ownership of all accountabilities and responsibilities. In this way, successful governance and management processes become institutionalized in the culture.

As technical assessment is reported (e.g., through performance metrics such as those established through process MEA01) to business process owners, they, in turn, report progress toward enterprise goals and mission priorities, using language, approaches and communications that are meaningful to executive management.

Figure 3.10 details the implementation considerations for COBIT Implementation Phase 7 and NIST CSF Lifecycle Management.

Figure 3.10—Implementation Considerations NIST CSF Lifestyle Management

Implementation Considerations	
COBIT Implementation Phase 7—How Do We Keep the Momentum Going? NIST CSF Lifecycle Management (not a CSF step)	
Description	
This phase enables the team to determine whether the program delivered against expectations.	
Objectives	
• Assess the results and experience gained from the program. • Record and share any lessons learned. • Ensure that new, required actions drive further iterations of the life cycle. • Continually monitor performance and ensure results are regularly reported. • Drive commitment and ownership of all accountabilities and responsibilities.	
Key Tasks and Activities	
Monitor and evaluate:	
• Identify new governance objectives and requirements. • Gather feedback. • Measure and report actual results against originally established project measures. • Perform a facilitated project review process. • Look for additional high-impact, low-cost opportunities to further improve EGIT. • Identify lessons learned. • Communicate requirements for further improvements.	
Sustain:	
• Provide conscious reinforcement and ongoing communication. • Confirm conformance to objectives and requirements. • Continually monitor the effectiveness of the change(s). • Implement corrective action plans where required. • Provide feedback on performance and publicize successes. • Build on lessons learned. • Share knowledge from the initiative to the broader enterprise.	
Review program effectiveness:	
• At program closure, ensure that a program review takes place and approve conclusions. • Review program effectiveness.	
Inputs	**Outputs**
• Updated project and program scorecards • Change-effectiveness measures (both business and perception measures) • Report explaining scorecard results • Postimplementation review report • Performance reports • Business and IT strategy • New triggers such as new regulatory requirements	• Recommendations for further EGIT activities after a period of normalization • Stakeholder satisfaction survey • Documented success stories and lessons learned • Ongoing communication plan • Performance reward scheme
COBIT Design Workflow (no relationship)	
Available Resources	
• *COBIT® 2019 Framework: Governance and Management Objectives* (EDM01, APO01, BAI08, MEA01), *www.isaca.org/cobit* • ISACA supporting products as currently listed at *www.isaca.org*	

Additional Practice Guidance from COBIT 2019	
COBIT 2019 Practice	Description
EDM01.03	**Monitor the governance system.** Monitor the effectiveness and performance of the enterprise's governance of I&T. Assess whether the governance system and implemented mechanisms (including structures, principles and processes) are operating effectively and provide appropriate oversight of I&T to enable value creation.
EDM02.02	**Evaluate value optimization.** Continually evaluate the portfolio of I&T-enabled investments, services and assets to determine the likelihood of achieving enterprise objectives and delivering value. Identify and evaluate any changes in direction to management that will optimize value creation.
EDM02.04	**Monitor value optimization.** Monitor key goals and metrics to determine whether the enterprise receives expected value and benefit from I&T-enabled investments and services. Identify significant issues and consider corrective actions.
EDM03.03	**Monitor risk management**. Monitor the key goals and metrics of the risk management processes. Determine how deviations or problems will be identified, tracked and reported for remediation.
EDM04.03	**Monitor resource management**. Monitor the key goals and metrics of the resource management processes. Determine how deviations or problems will be identified, tracked and reported for remediation.
EDM05.03	**Monitor stakeholder engagement**. Monitor stakeholder engagement levels and the effectiveness of stakeholder communication. Assess mechanisms for ensuring accuracy, reliability and effectiveness, and ascertain whether the requirements of different stakeholders in terms of reporting and communication are met.
APO04.03	**Monitor and scan the technology environment.** Set up a technology watch process to perform systematic monitoring and scanning of the enterprise's external environment to identify emerging technologies that have the potential to create value (e.g., by realizing the enterprise strategy, optimizing costs, avoiding obsolescence, and better enabling enterprise and I&T processes). Monitor the marketplace, competitive landscape, industry sectors, and legal and regulatory trends to be able to analyze emerging technologies or innovation ideas in the enterprise context.
APO04.04	**Assess the potential of emerging technologies and innovative ideas.** Analyze identified emerging technologies and/or other I&T innovative suggestions to understand their business potential. Work with stakeholders to validate assumptions on the potential of new technologies and innovation.
APO04.05	**Recommend appropriate further initiatives**. Evaluate and monitor the results of proof-of-concept initiatives and, if favorable, generate recommendations for further initiatives. Gain stakeholder support.
APO04.06	**Monitor the implementation and use of innovation**. Monitor the implementation and use of emerging technologies and innovations during adoption, integration and for the full economic life cycle to ensure that the promised benefits are realized and to identify lessons learned.
APO05.03	**Monitor, optimize and report on investment portfolio performance**. On a regular basis, monitor and optimize the performance of the investment portfolio and individual programs throughout the entire investment life cycle. Ensure continuous follow-up on the alignment of the portfolio with I&T strategy.
APO05.04	**Maintain portfolios**. Maintain portfolios of investment programs and projects, I&T products and services, and I&T assets.
APO05.05	**Manage benefits achievement**. Monitor the benefits of providing and maintaining appropriate I&T products, services and capabilities, based on the agreed and current business case.
APO07.05	**Plan and track the usage of IT and business human resources**. Understand and track the current and future demand for business and IT human resources with responsibilities for enterprise I&T. Identify shortfalls and provide input into sourcing plans, enterprise and IT recruitment processes, and business and IT recruitment processes.
APO07.06	**Manage contract staff**. Ensure that consultants and contract personnel who support the enterprise with I&T skills know and comply with the organization's policies and meet agreed contractual requirements.
APO08.05	**Provide input to the continual improvement of services.** Continually improve and evolve I&T-enabled services and service delivery to the enterprise to align with changing enterprise objectives and technology requirements.
APO09.04	**Monitor and report service levels.** Monitor service levels, report on achievements and identify trends. Provide the appropriate management information to aid performance management.
APO09.05	**Review service agreements and contracts.** Conduct periodic reviews of the service agreements and revise when needed.

Additional Practice Guidance from COBIT 2019 *(cont.)*	
COBIT 2019 Practice	**Description**
APO10.03	**Manage vendor relationships and contracts.** Formalize and manage the supplier relationship for each supplier. Manage, maintain and monitor contracts and service delivery. Ensure that new or changed contracts conform to enterprise standards and legal and regulatory requirements. Deal with contractual disputes.
APO10.04	**Manage vendor risk**. Identify and manage risk relating to vendors' ability to continually provide secure, efficient and effective service delivery. This also includes the subcontractors or upstream vendors that are relevant in the service delivery of the direct vendor.
APO10.05	**Monitor vendor performance and compliance**. Periodically review overall vendor performance, compliance to contract requirements and value for money. Address identified issues.
APO11.04	**Perform quality monitoring, control and reviews**. Monitor the quality of processes and services on an ongoing basis, in line with quality management standards. Define, plan and implement measurements to monitor customer satisfaction with quality as well as the value provided by the quality management system (QMS). The information gathered should be used by the process owner to improve quality.
APO11.05	**Maintain continuous improvement.** Maintain and regularly communicate an overall quality plan that promotes continuous improvement. The plan should define the need for, and benefits of, continuous improvement. Collect and analyze data about the quality management system (QMS) and improve its effectiveness. Correct nonconformities to prevent recurrence.
APO13.01	**Establish and maintain an information security management system (ISMS).** Establish and maintain an information security management system (ISMS) that provides a standard, formal and continuous approach to information security management, enabling secure technology and business processes that are aligned with business requirements.
APO13.02	**Define and manage an information security and privacy risk treatment plan.** Maintain an information security plan that describes how information security risk is to be managed and aligned with enterprise strategy and architecture. Ensure that recommendations for implementing security improvements are based on approved business cases, implemented as an integral part of services and solutions development, and operated as an integral part of business operation.
APO13.03	**Monitor and review the information security management system (ISMS).** Maintain and regularly communicate the need for, and benefits of, continuous improvement in information security. Collect and analyze data about the information security management system (ISMS), and improve its effectiveness. Correct nonconformities to prevent recurrence.
BAI01.06	**Monitor, control and report on the program outcomes.** Monitor and control performance against plan throughout the full economic life cycle of the investment, covering solution delivery at the program level and value/outcome at the enterprise level. Report performance to the program steering committee and the sponsors.
BAI01.08	**Manage program risk.** Eliminate or minimize specific risk associated with programs through a systematic process of planning, identifying, analyzing, responding to, monitoring and controlling the areas or events with the potential to cause unwanted change. Define and record any risk faced by program management.
BAI03.09	**Manage changes to requirements**. Track the status of individual requirements (including all rejected requirements) throughout the project life cycle. Manage the approval of changes to requirements.
BAI03.10	**Maintain solutions**. Develop and execute a plan for the maintenance of solution and infrastructure components. Include periodic reviews against business needs and operational requirements.
BAI04.04	**Monitor and review availability and capacity**. Monitor, measure, analyze, report and review availability, performance and capacity. Identify deviations from established baselines. Review trend analysis reports identifying any significant issues and variances. Initiate actions where necessary and ensure that all outstanding issues are addressed.
BAI05.07	**Sustain changes**. Sustain changes through effective training of new staff, ongoing communication campaigns, continued commitment of top management, monitoring of adoption and sharing of lessons learned across the enterprise.
BAI06 (all)	**Managed IT Changes.** Manage all changes in a controlled manner, including standard changes and emergency maintenance relating to business processes, applications and infrastructure. This includes change standards and procedures, impact assessment, prioritization and authorization, emergency changes, tracking, reporting, closure, and documentation.

Additional Practice Guidance from COBIT 2019 *(cont.)*	
COBIT 2019 Practice	**Description**
BAI07 (all)	**Managed IT Change Acceptance and Transitioning.** Formally accept and make operational new solutions. Include implementation planning, system and data conversion, acceptance testing, communication, release preparation, promotion to production of new or changed business processes and I&T services, early production support, and a post-implementation review.
BAI08 (all)	**Managed Knowledge.** Maintain the availability of relevant, current, validated and reliable knowledge and management information to support all process activities and to facilitate decision making related to the governance and management of enterprise I&T. Plan for the identification, gathering, organizing, maintaining, use and retirement of knowledge.
BAI10.03	**Maintain and control configuration items.** Maintain an up-to-date repository of configuration items (CIs) by populating any configuration changes.
BAI11.06	**Manage project risk.** Eliminate or minimize specific risk associated with projects through a systematic process of planning, identifying, analyzing, responding to, monitoring and controlling the areas or events with potential to cause unwanted change. Define and record any risk faced by project management.
BAI11.07	**Monitor and control projects**. Measure project performance against key project performance criteria such as schedule, quality, cost and risk. Identify any deviations from expected targets. Assess the impact of deviations on the project and overall program and report results to key stakeholders.
BAI11.08	**Manage project resources and work packages**. Manage project work packages by placing formal requirements on authorizing and accepting work packages and assigning and coordinating appropriate business and IT resources.
BAI10.04	**Produce status and configuration reports**. Define and produce configuration reports on status changes of configuration items.
BAI10.05	**Verify and review integrity of the configuration repository**. Periodically review the configuration repository and verify completeness and correctness against the desired target.
DSS01 (all)	**Managed Operations.** Coordinate and execute the activities and operational procedures required to deliver internal and outsourced I&T services. Include the execution of predefined standard operating procedures and the required monitoring activities.
DSS02 (all)	**Managed Service Requests and Incidents.** Provide timely and effective response to user requests and resolution of all types of incidents. Restore normal service; record and fulfil user requests; and record, investigate, diagnose, escalate and resolve incidents.
DSS03 (all)	**Managed Problems.** Identify and classify problems and their root causes. Provide timely resolution to prevent recurring incidents. Provide recommendations for improvements.
DSS04 (all)	**Managed Continuity.** Establish and maintain a plan to enable the business and IT organizations to respond to incidents and quickly adapt to disruptions. This will enable continued operations of critical business processes and required I&T services and maintain availability of resources, assets and information at a level acceptable to the enterprise.
MEA01.04	**Analyze and report performance.** Periodically review and report performance against targets. Use a method that provides a succinct all-around view of I&T performance and fits within the enterprise monitoring system.
MEA01.05	**Ensure the implementation of corrective actions.** Assist stakeholders in identifying, initiating and tracking corrective actions to address anomalies.
MEA02 (all)	**Managed System of Internal Control.** Continuously monitor and evaluate the control environment, including self-assessments and self-awareness. Enable management to identify control deficiencies and inefficiencies and to initiate improvement actions. Plan, organize and maintain standards for internal control assessment and process control effectiveness.

Chapter 4
Communicating Cybersecurity Requirements With Stakeholders

4.1 Introduction

The CSF stresses a common language that can be used to communicate requirements among interdependent stakeholders responsible for critical infrastructure products and services. Communication is especially important among stakeholders within the enterprise supply chain.

An important component of both the CSF and COBIT 2019 involves the governance and management of suppliers and business partners. A single business system may encompass dozens of external stakeholders and supply chain/service providers. Each of these stakeholders represents an opportunity to fulfill enterprise and alignment (or I&T-related) goals; they also add additional vulnerability and potential risk to be considered. Implementing the CSF using COBIT 2019 principles and processes helps to establish a common language for communicating stakeholder needs and requirements, and also helps enterprises address the specific risk inherent in extended networks of suppliers and partners.

Implementing the CSF using COBIT 2019 principles and processes helps to establish a common language for communicating stakeholder needs and requirements, and also helps enterprises address the specific risk inherent in extended networks of suppliers and partners.

4.2 Cyber Supply Chain Risk Management under CSF and COBIT 2019

CSF Version 1.1 provides detail on supply chain risk management (SCRM). Under the CSF, cyber SCRM addresses the cybersecurity effect that an enterprise has on its external partners and the (reciprocal) cybersecurity effect that third parties have on the enterprise.

Cyber SCRM activities include:

- Determining cybersecurity requirements for vendors
- Establishing cybersecurity requirements through formal agreements and contracts
- Communicating to vendors how cybersecurity requirements will be verified and validated
- Performing assessments to validate fulfillment of cybersecurity requirements
- Governing and managing these activities

This guidance is consistent with COBIT 2019 management objective APO10 *Managed vendors*, and specifically management practices APO10.04 *Manage vendor risk* and APO10.05 *Monitor vendor performance and compliance.*

COBIT positions I&T to be governed and managed holistically for the entire enterprise, supporting the primary organization and its supply chain partners. Many COBIT 2019 practices include vendor components, guided by APO10 *Managed vendors*. COBIT 2019 informs many aspects of vendor relations in the context of CSF implementation, including:

- Documentation of supplier management. Cooperative agreements provide an opportunity to record drivers, risk agreements and goals, using a subset of the processes in COBIT Implementation Phase 1 (section 3).
- Documentation of supplier/partner assessments. The Current Profile records the current state of cybersecurity risk management. Alignment around this CSF/COBIT model supports the COBIT principle of a single, integrated framework model to record and communicate goals and performance. The Current Profile may be shared with third-party providers to communicate current achievement ratings and ongoing actions, etc.

- Documentation of cybersecurity services management, assessment and evolution. The Target Profile records cybersecurity risk management requirements and goals and may be shared with third-party service providers to communicate expectations, recommended actions, etc. (as described in section 3, COBIT Implementation Phase 3). This model is helpful for conveying specific EGIT obligations—for example, to a cloud provider to which the enterprise exports data.

Harmonizing processes and communication—for both internal and external stakeholders—improves consistency and simplifies tracking/reporting. Common templates and communication practices help to consolidate an holistic approach to governance and management of I&T and ensure that goals are aligned and effective.

APPENDIX A

Framework Core

As described in section 2 of this publication, the Framework Core provides a set of activities to achieve specific cybersecurity outcomes and references key guidance to achieve those goals. The Core is not a checklist of actions to perform. It presents key cybersecurity outcomes identified by industry as helpful in managing cybersecurity risk. The Core comprises four elements: Functions, Categories, Subcategories and Informative References.

The following table represents the Framework Core as provided in Appendix A of the NIST *Framework for Improving Critical Infrastructure Cybersecurity*, Version 1.1, mapped to COBIT® 2019 Informative References.

As cybersecurity threats and technology continually evolve, existing standards and requirements change to remain relevant and effective. NIST established an Online Informative References (OLIR) Catalog (*https://www.nist.gov/cyberframework/informative-references/informative-reference-catalog*) that builds on published Informative References in the Framework Core. The OLIR catalog reflects verified CSF V1.1 mapping of Informative References subsequent to the release of V1.1. The dynamic nature of the catalog requires that practitioners periodically review it for additions and changes to Informative References.

Function	Category	Subcategory	Informative References
IDENTIFY (ID)	**Asset Management (ID.AM):** The data, personnel, devices, systems, and facilities that enable the organization to achieve business purposes are identified and managed consistent with their relative importance to business objectives and the organization's risk strategy.	**ID.AM-1:** Physical devices and systems within the organization are inventoried.	• BAI09.01, BAI09.02
		ID.AM-2: Software platforms and applications within the organization are inventoried.	• BAI09.01, BAI09.02, BAI09.05
		ID.AM-3: Organizational communication and data flows are mapped.	• APO14.08 • DSS05.02
		ID.AM-4: External information systems are catalogued.	• APO02.02, APO10.01, APO10.04 • DSS01.02
		ID.AM-5: Resources (e.g., hardware, devices, data, and software) are prioritized based on their classification, criticality, and business value.	• APO03.03, APO03.04, APO12.01 • BAI04.02, BAI09.02
		ID.AM-6: Cybersecurity roles and responsibilities for the entire workforce and third-party stakeholders (e.g., suppliers, customers, partners) are established.	• APO01.02, APO01.05, APO07.06, APO13.01 • DSS06.03
	Business Environment (ID.BE): The organization's mission, objectives, stakeholders, and activities are understood and prioritized; this information is used to inform cybersecurity roles, responsibilities, and risk management decisions.	**ID.BE-1:** The organization's role in the supply chain is identified and communicated.	• APO08.01, APO08.04, APO08.05, APO10.03, APO10.04, APO10.05
		ID.BE-2: The organization's place in critical infrastructure and its industry sector is identified and communicated.	• APO02.06, APO03.01
		ID.BE-3: Priorities for organizational mission, objectives, and activities are established and communicated.	• EDM02.03 • APO02.01, APO02.06, APO03.01

Function	Category	Subcategory	Informative References
IDENTIFY (ID)	**Business Environment (ID.BE)** *(cont.)*	**ID.BE-4:** Dependencies and critical functions for delivery of critical services are established.	• APO10.01 • BAI04.02, BAI09.02
		ID.BE-5: Resilience requirements to support delivery of critical services are established.	• BAI03.02 • DSS04.02
	Governance (ID.GV): The policies, procedures, and processes to manage and monitor the organization's regulatory, legal, risk, environmental, and operational requirements are understood and inform the management of cybersecurity risk.	**ID.GV-1:** Organizational information security policy is established and communicated.	• EDM01.02 • APO01.02, APO01.03, APO13.01
		ID.GV-2 Cybersecurity roles and responsibilities are coordinated and aligned with internal roles and external partners.	• APO01.02, APO01.05, APO10.03, APO13.02 • DSS05.04
		ID.GV-3: Legal and regulatory requirements regarding cybersecurity, including privacy and civil liberties obligations, are understood and managed.	• BAI02.01 • MEA03.01, MEA03.03
		ID.GV-4: Governance and risk management processes address cybersecurity risks.	• EDM03.02 • APO12.02, APO12.05 • DSS04.02
	Risk Assessment (ID.RA): The organization understands the cybersecurity risk to organizational operations (including mission, functions, image, or reputation), organizational assets, and individuals.	**ID.RA-1:** Asset vulnerabilities are identified and documented.	• APO12.01, APO12.02, APO12.03, APO12.04 • DSS05.01, DSS05.02
		ID.RA-2: Threat and vulnerability information is received from information sharing forums and sources.	• BAI08.01, BAI08.02, BAI08.03
		ID.RA-3: Threats, both internal and external, are identified and documented.	• APO12.01, APO12.02, APO12.03, APO12.04
		ID.RA-4: Potential business impacts and likelihoods are identified.	• DSS04.02
		ID.RA-5: Threats, vulnerabilities, likelihoods, and impacts are used to determine risk.	• APO12.02
		ID.RA-6: Risk responses are identified and prioritized.	• APO12.05, APO13.02
	Risk Management Strategy (ID.RM): The organization's priorities, constraints, risk tolerances, and assumptions are established and used to support operational risk decisions.	**ID.RM-1:** Risk management processes are established, managed, and agreed to by organizational stakeholders.	• APO12.04, APO12.05, APO13.02 • BAI02.03, BAI04.02
		ID.RM-2: Organizational risk tolerance is determined and clearly expressed.	• EDM03.01 • APO12.06
		ID.RM-3: The organization's determination of risk tolerance is informed by its role in critical infrastructure and sector specific risk analysis.	• APO12.02

Function	Category	Subcategory	Informative References
IDENTIFY (ID)	**Supply Chain Risk Management (ID.SC):** The organization's priorities, constraints, risk tolerances, and assumptions are established and used to support risk decisions associated with managing supply chain risk. The organization has established and implemented the processes to identify, assess and manage supply chain risks.	**ID.SC-1:** Cyber supply chain risk management processes are identified, established, assessed, managed, and agreed to by organizational stakeholders.	• APO10.01, APO10.04, APO12.04, APO12.05, APO13.02 • BAI01.03, BAI02.03, BAI04.02
		ID.SC-2: Suppliers and third party partners of information systems, components, and services are identified, prioritized, and assessed using a cyber supply chain risk assessment process.	• APO10.01, APO10.02, APO10.04, APO10.05, APO12.01, APO12.02, APO12.03, APO12.04, APO12.05, APO12.06, APO13.02 • BAI02.03
		ID.SC-3: Contracts with suppliers and third-party partners are used to implement appropriate measures designed to meet the objectives of an organization's cybersecurity program and Cyber Supply Chain Risk Management Plan.	• APO10.01, APO10.02, APO10.03, APO10.04, APO10.05
		ID.SC-4: Suppliers and third-party partners are routinely assessed using audits, test results, or other forms of evaluations to confirm they are meeting their contractual obligations.	• APO10.01, APO10.03, APO10.04, APO10.05 • MEA01.01, MEA01.02, MEA01.03, MEA01.04, MEA02.01, MEA02.02, MEA03.03, MEA03.04, MEA04.07
		ID.SC-5: Response and recovery planning and testing are conducted with suppliers and third-party providers.	• DSS04.04
PROTECT (PR)	**Access Control (PR.AC):** Access to assets and associated facilities is limited to authorized users, processes, or devices, and to authorized activities and transactions.	**PR.AC-1:** Identities and credentials are managed for authorized devices and users.	• DSS05.04, DSS06.03
		PR.AC-2: Physical access to assets is managed and protected.	• DSS01.04, DSS05.05
		PR.AC-3: Remote access is managed.	• APO13.01 • BAI09.02, BAI09.03 • DSS01.04, DSS05.03
		PR.AC-4: Access permissions are managed, incorporating the principles of least privilege and separation of duties.	• DSS05.04
		PR.AC-5: Network integrity is protected, incorporating network segregation where appropriate.	• DSS05.02, DSS05.03
		PR.AC-6: Identities are proofed and bound to credentials and asserted in interactions.	• DSS05.04, DSS05.05, DSS05.07, DSS06.03
		PR.AC-7: Users, devices, and other assets are authenticated (e.g., single-factor, multi-factor) commensurate with the risk of the transaction (e.g., individuals' security and privacy risks and other organizational risks).	• DSS05.04, DSS06.03

Function	Category	Subcategory	Informative References
PROTECT (PR)	**Awareness and Training (PR.AT):** The organization's personnel and partners are provided cybersecurity awareness education and are adequately trained to perform their information security-related duties and responsibilities consistent with related policies, procedures, and agreements.	**PR.AT-1:** All users are informed and trained.	• APO07.03 • BAI05.07
		PR.AT-2: Privileged users understand roles and responsibilities.	• APO07.02 • DSS05.04, DSS06.03
		PR.AT-3: Third-party stakeholders (e.g., suppliers, customers, partners) understand roles and responsibilities.	• APO07.03, APO07.06, APO10.04, APO10.05
		PR.AT-4: Senior executives understand roles and responsibilities.	• EDM01.02 • APO01.02, APO07.03
		PR.AT-5: Physical and information security personnel understand roles and responsibilities.	• APO07.03 • DSS06.03
	Data Security (PR.DS): Information and records (data) are managed consistent with the organization's risk strategy to protect the confidentiality, integrity, and availability of information.	**PR.DS-1:** Data-at-rest is protected.	• APO01.07, APO14.04, APO14.05, APO14.08 • BAI02.01 • DSS04.07, DSS05.03, DSS06.06
		PR.DS-2: Data-in-transit is protected.	• APO01.07, APO14.04, APO14.05, APO14.08 • DSS05.02, DSS06.06
		PR.DS-3: Assets are formally managed throughout removal, transfers, and disposition.	• BAI09.03
		PR.DS-4: Adequate capacity to ensure availability is maintained.	• APO13.01 • BAI04.04, BAI04.05
		PR.DS-5: Protections against data leaks are implemented.	• APO01.07, APO13.02 • DSS05.04, DSS05.07, DSS06.02
		PR.DS-6: Integrity checking mechanisms are used to verify software, firmware, and information integrity.	• APO01.07 • BAI03.05, BAI06.01 • DSS06.02
		PR.DS-7: The development and testing environment(s) are separate from the production environment.	• BAI03.07, BAI03.08, BAI07.04
		PR.DS-8: Integrity checking mechanisms are used to verify hardware integrity.	• BAI03.05
	Information Protection Processes and Procedures (PR.IP): Security policies (that address purpose, scope, roles, responsibilities, management commitment, and coordination among organizational entities), processes, and procedures are maintained and used to manage protection of information systems and assets.	**PR.IP-1:** A baseline configuration of information technology/industrial control systems is created and maintained.	• BAI10.01, BAI10.02, BAI10.03, BAI10.05
		PR.IP-2: A System Development Life Cycle to manage systems is implemented.	• APO13.01 • BAI03.01, BAI03.02, BAI03.03
		PR.IP-3: Configuration change control processes are in place.	• BAI01.06, BAI06.01
		PR.IP-4: Backups of information are conducted, maintained, and tested periodically.	• APO13.01 • BAI07.02 • DSS01.01, DSS04.07

Function	Category	Subcategory	Informative References
PROTECT (PR)	**Information Protection Processes and Procedures (PR.IP)** *(cont.)*	**PR.IP-5:** Policy and regulations regarding the physical operating environment for organizational asssets are met.	• DSS01.04, DSS05.05
		PR.IP-6: Data is destroyed according to policy.	• APO14.08 • BAI09.03 • DSS05.06
		PR.IP-7: Protection processes are continuously improved.	• APO12.06 • DSS04.05
		PR.IP-8: Effectiveness of protection technologies is shared with appropriate parties.	• BAI08.03 • DSS03.04
		PR.IP-9: Response plans (Incident Response and Business Continuity) and recovery plans (Incident Recovery and Disaster Recovery) are in place and managed.	• APO12.06 • DSS04.02, DSS04.03
		PR.IP-10: Response and recovery plans are tested.	• DSS04.04
		PR.IP-11: Cybersecurity is included in human resources practices (e.g., deprovisioning, personnel screening).	• APO07.01, APO07.02, APO07.03, APO07.04, APO07.05 • DSS05.04, DSS05.05, DSS06.03
		PR.IP-12: A vulnerability management plan is developed and implemented.	• BAI03.10 • DSS05.01, DSS05.02, DSS05.07
	Maintenance (PR.MA): Maintenance and repairs of industrial control and information system components is performed consistent with policies and procedures.	**PR.MA-1:** Maintenance and repair of organizational assets is performed and logged in a timely manner, with approved and controlled tools.	• BAI03.10, BAI09.02, BAI09.03 • DSS01.05
		PR.MA-2: Remote maintenance of organizational assets is approved, logged, and performed in a manner that prevents unauthorized access.	• DSS05.04
	Protective Technology (PR.PT): Technical security solutions are managed to ensure the security and resilience of systems and assets, consistent with related policies, procedures, and agreements.	**PR.PT-1:** Audit/log records are determined, documented, implemented, and reviewed in accordance with policy.	• APO11.04 • BAI03.05 • DSS05.04, DSS05.07 • MEA02.01
		PR.PT-2: Removable media is protected and its use restricted according to policy.	• APO13.01 • DSS05.02, DSS05.06
		PR.PT-3: Access to systems and assets is controlled, incorporating the principle of least functionality.	• BAI10.01, BAI10.02, BAI10.03 • DSS05.02, DSS05.05, DSS06.06
		PR.PT-4: Communications and control networks are protected.	• APO13.01 • DSS05.02
		PR.PT-5: Mechanisms (e.g., failsafe, load balancing, hot swap) are implemented to achieve resilience requirements in normal and adverse situations.	• BAI04.01, BAI04.02, BAI04.03, BAI04.04, BAI04.05, BAI09.02 • DSS01.05, DSS04.02

Function	Category	Subcategory	Informative References
DETECT (DT)	**Anomalies and Events (DE.AE):** Anomalous activity is detected in a timely manner and the potential impact of events is understood.	**DE.AE-1:** A baseline of network operations and expected data flows for users and systems is established and managed.	• DSS03.01
		DE.AE-2: Detected events are analyzed to understand attack targets and methods.	• DSS05.07
		DE.AE-3: Event data are aggregated and correlated from multiple sources and sensors.	• APO14.08, APO14.09 • BAI08.01 • DSS01.03, DSS05.07
		DE.AE-4: Impact of events is determined.	• APO12.06 • BAI04.02 • DSS03.01
		DE.AE-5: Incident alert thresholds are established.	• APO12.06 • DSS01.03, DSS03.01
	Security Continuous Monitoring (DE.CM): The information system and assets are monitored at discrete intervals to identify cybersecurity events and verify the effectiveness of protective measures.	**DE.CM-1:** The network is monitored to detect potential cybersecurity events.	• DSS01.03, DSS03.05, DSS05.07
		DE.CM-2: The physical environment is monitored to detect potential cybersecurity events.	• DSS01.04, DSS01.05
		DE.CM-3: Personnel activity is monitored to detect potential cybersecurity events.	• DSS05.07
		DE.CM-4: Malicious code is detected.	• DSS05.01
		DE.CM-5: Unauthorized mobile code is detected.	• DSS05.01
		DE.CM-6: External service provider activity is monitored to detect potential cybersecurity events.	• APO07.06, APO10.05
		DE.CM-7: Monitoring for unauthorized personnel, connections, devices, and software is performed.	• DSS05.02, DSS05.04, DSS05.05
		DE.CM-8: Vulnerability scans are performed.	• DSS05.07
	Detection Processes (DE.DP): Detection processes and procedures are maintained and tested to ensure timely and adequate awareness of anomalous events.	**DE.DP-1:** Roles and responsibilities for detection are well defined to ensure accountability.	• APO01.05 • DSS06.03 • MEA04.06
		DE.DP-2: Detection activities comply with all applicable requirements.	• DSS06.01 • MEA03.03, MEA03.04
		DE.DP-3: Detection processes are tested.	• APO13.03 • BAI03.07, BAI03.08, BAI07.05 • DSS05.02
		DE.DP-4: Event detection information is communicated to appropriate parties.	• APO08.04, APO12.06 • DSS03.02, DSS05.07
		DE.DP-5: Detection processes are continuously improved.	• APO11.05, APO12.06, APO13.03 • DSS04.05

Function	Category	Subcategory	Informative References
RESPOND (RS)	**Analysis (RS.AN):** Analysis is conducted to ensure adequate response and support recovery activities.	**RS.AN-1:** Notifications from detection systems are investigated.	• DSS02.04, DSS02.07
		RS.AN-2: The impact of the incident is understood.	• APO12.06 • DSS02.02, DSS03.01, DSS03.02
		RS.AN-3: Forensics are performed.	• APO12.06 • DSS03.02, DSS05.07
		RS.AN-4: Incidents are categorized consistent with response plans.	• DSS02.02
		RS.AN-5: Processes are established to receive, analyze and respond to vulnerabilities disclosed to the organization from internal and external sources (e.g. internal testing, security bulletins, or security researchers).	• EDM02.04, EDM03.02 • BAI08.03 • DSS05.07
	Communications (RS.CO): Response activities are coordinated with internal and external stakeholders, as appropriate, to include external support from law enforcement agencies.	**RS.CO-1:** Personnel know their roles and order of operations when a response is needed.	• EDM03.02 • APO01.02, APO01.05
		RS.CO-2: Events are reported consistent with established criteria.	• DSS01.03
		RS.CO-3: Information is shared consistent with response plans.	• DSS03.03, DSS03.04
		RS.CO-4: Coordination with stakeholders occurs consistent with response plans.	• DSS03.04, DSS03.05
		RS.CO-5: Voluntary information sharing occurs with external stakeholders to achieve broader cybersecurity situational awareness.	• BAI08.03
	Improvements (RS.IM): Organizational response activities are improved by incorporating lessons learned from current and previous detection/response activities.	**RS.IM-1:** Response plans incorporate lessons learned.	• BAI01.09
		RS.IM-2: Response strategies are updated.	• BAI01.09 • DSS04.08
	Mitigation (RS.MI): Activities are performed to prevent expansion of an event, mitigate its effects, and eradicate the incident.	**RS.MI-1:** Incidents are contained.	• APO12.06 • DSS05.03, DSS05.07
		RS.MI-2: Incidents are mitigated.	• APO12.06 • DSS05.03, DSS05.07
		RS.MI-3: Newly identified vulnerabilities are mitigated or documented as accepted risks.	• APO12.06 • DSS05.07
	Response Planning (RS.RP): Response processes and procedures are executed and maintained, to ensure timely response to detected cybersecurity events.	**RS.RP-1:** Response plan is executed during or after an event.	• APO12.06 • DSS02.05, DSS03.04

Function	Category	Subcategory	Informative References
RECOVER (RC)	**Communications (RC.CO):** Restoration activities are coordinated with internal and external parties, such as coordinating centers, Internet Service Providers, owners of attacking systems, victims, other CSIRTs, and vendors.	**RC.CO-1:** Public relations are managed.	• EDM03.02
		RC.CO-2: Reputation after an event is repaired.	• APO12.06 • MEA03.02
		RC.CO-3: Recovery activities are communicated to internal stakeholders and executive and management teams.	• APO12.06
	Improvements (RC.IM): Recovery planning and processes are improved by incorporating lessons learned into future activities.	**RC.IM-1:** Recovery plans incorporate lessons learned.	• APO12.06 • BAI05.07 • DSS04.08
		RC.IM-2: Recovery strategies are updated.	• APO12.06 • BAI07.08 • DSS04.08
	(RC.RP): Recovery processes and procedures are executed and maintained to ensure timely restoration of systems or assets affected by cybersecurity events.	**RC.RP-1:** Recovery plan is executed during or after an event.	• APO12.06 • DSS02.05, DSS03.04

Source: NIST, *Framework for Improving Critical Infrastructure Cybersecurity*, Appendix A, USA, 2018

APPENDIX B

Detailed Template for Framework Profile

As discussed in section 2 of this publication, the Framework Profile allows an enterprise to track the current state of its cybersecurity program and also identify characteristics of the target state. This appendix provides example templates for the enterprise to construct Current and Target Profiles.

Profile Metadata

The Profile metadata table (**figure B.1**) is used to capture information regarding the organization, business unit or system(s) that are represented in the Profile. This information is typically collected in Phases 1 and 2 of the CSF implementation process.

Figure B.1—Profile Metadata Table

Organization	
Organization/business unit/sector/subsidiary (if applicable)	
Organization current Profile scope	
Business requirements	**1.** **2.** **3.** **4.**
Risk considerations	
Risk appetite decisions	

Target Profile

The target-state profile captures the desired end state of the cybersecurity program. The Target Profile should identify the protections and capabilities that are required to mitigate threats to the enterprise. This risk-based approach ensures that all areas of the CSF are addressed, with a focus on those areas most likely to be attacked. Target Profile data points are shown in **figure B.2**.

Figure B.2—Target Profile Data Points

Topic	Required information
Function	Applicable Framework Function
Category	Applicable Framework Category
Subcategory	Applicable Framework Subcategory
Relevant COBIT processes	COBIT 2019 Informative Reference identifying the enterprise practices required to attain goals of the CSF Subcategory
Implementation status	Achievement rating (i.e., Not achieved, Partially achieved, Largely achieved, Fully achieved) based on the implementation of existing policies and procedures
Organizational practices	Practice, policy or procedure identified through the risk-based assessment of enterprise needs to mitigate risk related to the Category
Comments/evidence	Narrative describing how the achievement rating was determined and any established, ongoing actions toward the goal of the Subcategory
Recommended actions	Actions required to achieve target-state goals
Resources required	Enterprise resources required to complete the recommended actions

Gap Analysis

For each of the subcategories in the Target Profile, practitioners should consider the difference between the target level of achievement and the current level. Understanding gaps between current and target enterprise policies and practices highlights opportunities for improvement. Understanding the relative impact on risk helps to establish priority, scheduling and resource allocation. Based on information from the gap analysis, practitioners may conduct activity planning. Appendix C of this publication presents detailed guidance for activity planning.

APPENDIX C
Action Planning

To achieve desired outcomes as described in the CSF and attain stakeholder goals identified in implementation Step 1, a comprehensive action plan is necessary. As part of the planning process, implementers should identify proper authorities who will review, approve and track activities and actions. Business/mission drivers should inform and support the actions.

Linking actions to enterprise and technical goals (as described in the COBIT 2019 goals cascade and documented in implementation Step 1) ensures that actions are able to be assessed and are prioritized to achieve the necessary value for the enterprise. Priorities and associated actions may be reviewed and adjusted through periodic checkpoint meetings, such as quarterly briefing program management reviews and security training exercises. A list of action plan data points is shown in **figure C.1**.

Specific considerations for action planning may include the following:

- Are there sector-specific action plan processes?
- Who is responsible for defining actions within the plan?
- How often will action plans be reviewed and updated? By whom?
- What specific governance and management processes apply within the industry?
- What are the advantages to achieving a higher/lower tier?
- What are the disadvantages to achieving a higher/lower tier?
- What regulatory guidance is available to help select the appropriate tier for the enterprise?
- What agencies, groups or consortia exist to support enterprise compliance and security programs?
- How is feedback captured and disseminated throughout the organization?

Figure C.1—Action Plan Data Points

Action Plan Detail	Description
Action identifier	• Unique identifier assigned to a specific action
Priority	• Enterprise-defined priority for completing the action (e.g., high, medium, low or 1 through 6)
Assumptions/constraints	• Enterprise-defined factors that may impact the ability to complete the action
Rationale	• Justification or basis for defining the action, including links to Profile(s) or regulatory requirements, when available
Specific action	• The discrete, outcome-based action to be completed
Resources required	• The enterprise resources required to complete the action
Schedule/milestones	• Key milestones and schedules assigned to the specific action
Status	• A stop light indicator (e.g., green, amber, red or other current, enterprise-approved status indicator) to signify the status of the action and identify issues that may cause a scheduled milestone to be missed
Prerequisites/dependencies	• Actions or enterprise factors that must be completed prior to this action being complete
Action assignee	• Point of contact assigned responsibility for tracking and ensuring that the action is completed
Stakeholder roles	• Internal and external enterprise stakeholder of the action

Page intentionally left blank

APPENDIX D

Considerations for Critical Infrastructure Sectors

NIST developed the CSF under EO 13636 to support the critical infrastructure community. Enterprises that identify with one of the 16 critical infrastructure sectors in **figure 1.2**—including any enterprise that supports those primary enterprises—should find the following considerations helpful when implementing the CSF.

Role Identification

From the chairperson of the board to operators at a nuclear reactor or linemen for an electric utility (for example), roles vary widely among critical infrastructure providers. The CSF generally classifies roles into three categories, as described in section 2 of this publication. Practitioners should carefully consider roles and titles, and refer consistently to those titles in planning, operations and monitoring documents. Doing so will facilitate education of stakeholders and employees, expedite implementation of cybersecurity activities, and minimize confusion over CSF role identification and local usage.

Implementation Scope

The optimal scope for CSF implementation varies with each enterprise. Some enterprises may prefer an incremental, exploratory approach, and thus apply the CSF to a subsidiary organization to gain experience; others may apply it to the entire enterprise at once. Such decisions are typically based on enterprise business needs and budgets.

Practitioners should determine whether any regulatory drivers affect scope of implementation. For example, the Health Insurance Portability and Accountability Act (HIPAA) describes specific objectives for "meaningful use" of certified electronic health record technology. Jurisdictional considerations may affect scope decisions—regulations in one country may be quite different from those in another part of the world. These external drivers may influence the goals considered and actions taken to improve cybersecurity.

Risk Considerations

Determination of enterprise risk architecture is an essential element of implementation Step 1, because many subsequent activities are designed to balance benefits with optimal risk levels and resource use.

Many critical information sectors are subject to external drivers that impact risk decisions. In the financial sector, for example, many factors influence consideration of acceptable risk. Documenting risk factors and considerations during Step 1 supports subsequent steps and ensures that stakeholder goals are attained and tracked, consistent with regulatory management and reporting requirements.

Quality Management

Quality management tracks closely with effective cybersecurity practice. COBIT 2019 process APO11 *Managed quality* describes the use and maintenance of a Quality Management System (QMS). Management practice APO11.01 states:

> *Establish and maintain a quality management system (QMS) that provides a standard, formal and continuous approach to quality management of information. The QMS should enable technology and business processes to align with business requirements and enterprise quality management.*

Applying the APO11 management practices helps the enterprise define and manage quality standards, practices and procedures, consistent with prioritization and risk decisions agreed on in the CSF implementation steps described earlier in this publication. Focusing quality management on customers and stakeholder goals (as established in phases 1 and 2) and integrating those quality management processes into the action plan, will help ensure alignment with mission. Performing quality monitoring, control and reviews helps ensure that enterprise processes and technology deliver value to the business, improve continuously and remain transparent for stakeholders.

Critical infrastructure providers may have additional QMS requirements for enterprise systems. The relevant goals for management of such a QMS should be considered when developing Profiles and determining actions. In this context, readers may be guided by standards in the ISO 9000 family, including:

- ISO 9001:2015—*Quality management systems — Requirements*
- ISO 9000:2015—*Quality management systems — Fundamentals and vocabulary*
- ISO 9004:2018—*Quality management — Quality of an organization — Guidance to achieve sustained success*
- ISO 19011:2018—*Guidelines for auditing management systems*

Threat and Vulnerability Information

Members of the critical infrastructure community are particular targets of cybersecurity threats, often through innovative attack vectors. US practitioners are encouraged to work with groups such as Information Sharing and Analysis Centers (ISACs) and the Department of Homeland Security, including the US Computer Emergency Readiness Team (CERT). InfraGard®, a partnership between the US Federal Bureau of Investigation (FBI) and the private sector, is also helpful. Its members represent businesses, academic institutions, state and local law enforcement agencies, and other participants dedicated to sharing information and intelligence to prevent hostile acts.

The National Council of ISACs (NCI) facilitates awareness and understanding of the enterprise threat and vulnerability landscape. NCI exists to advance the physical, and cybersecurity of, the critical infrastructures of North America by maximizing data exchange between ISACs, government, and private sector partners.

The Industrial Control System ISAC (ICS-ISAC) established a project known as the Situational Awareness Reference Architecture (SARA). The objective of SARA is to compile and publish an applied guide to the processes, practices, standards and technologies which facilities and others can use to establish situational awareness.

Enterprises should determine the conditions under which a vulnerability may be addressed. For example, some critical systems cannot necessarily be shut down just to apply an important patch, so mitigating controls should be identified to support enterprise goals for availability and security. These considerations apply to all people, processes and technology (as described in section 1) that enable business functions.

To support the growing need for information sharing, NIST published Special Publication 800-150,[23] which provides high-level guidance on forming, joining, and effective participation in information sharing communities.

Supply Chain Risk Management

Similarly, NIST promotes increased adoption of standards for supply chain risk management. NIST indicates that the:

> *adoption of supply chain risk management standards, practices and guidelines requires greater awareness and understanding of the risk associated with the time-sensitive interdependencies throughout the supply chain, including in and between critical infrastructure sectors/subsectors. This understanding is vital to enable organizations to assess their risk, prioritize, and allow for timely mitigation.*

[23] NIST, *NIST Guide to Cyber Threat Information Sharing,* October 2016, USA, *https://nvlpubs.nist.gov/nistpubs/SpecialPublications/NIST.SP.800-150.pdf*

CSF implementers are encouraged to include supply chain risk as a subset of their broad risk assessment and risk management activities. More information about supply chain risk management is available from the NIST Computer Security Division.

NIST deemed supply chain risk management critical enough to include in the Framework Core, adding it as a separate Category in Version 1.1.

Current and Target Profiles

During numerous CSF development workshops, NIST pointed out the potential that leaders of individual sectors (e.g., sector supporting agencies, sector councils, participating enterprises) would provide specific guidance on creation and maintenance of Current and Target Profiles. Such guidance might include:

- Mapping from the CSF Framework Core to compliance frameworks
- Criteria for determining thresholds described in **figure 3.3**
- Recommendations regarding Framework Core subcategories

Framework Next Steps

In announcing the launch of CSF 1.0, former Special Assistant to the US President and the US Cybersecurity Coordinator, Michael Daniel, made three requests that are still relevant today:[24]

> *We need you to kick the tires. We need organizations to begin using the Framework and see how well it can work for different sizes and types of organizations.*
>
> *We need your feedback to make the Framework better. We need you to share your experience with us on how using the Framework worked—or didn't work—for your organization. Feedback is essential to improving the Framework and making it better in future versions.*
>
> *In short, we need your continued engagement. The Framework is intended to be a living document. We need your collective experience and knowledge to make it better over time.*

CSF will continue to evolve, driven primarily by continued use and feedback. ISACA encourages all who implement CSF 1.1 to provide feedback to the CSF community. Doing so will advance the CSF's goal of improving cybersecurity risk management.

For practitioners worldwide, ISACA membership can be particularly helpful to achieve that latter, overarching goal—and safeguard enterprises around the globe.

[24] See Daniel, M.; The White House: President Barack Obama, Blog, "Launch of the Cybersecurity Framework — What's Next?," 18 February 2014, *https://obamawhitehouse.archives.gov/blog/2014/02/18/launch-cybersecurity-framework-what-s-next.*

Page intentionally left blank

9781604208153